T0364955

Art of the Grimoire

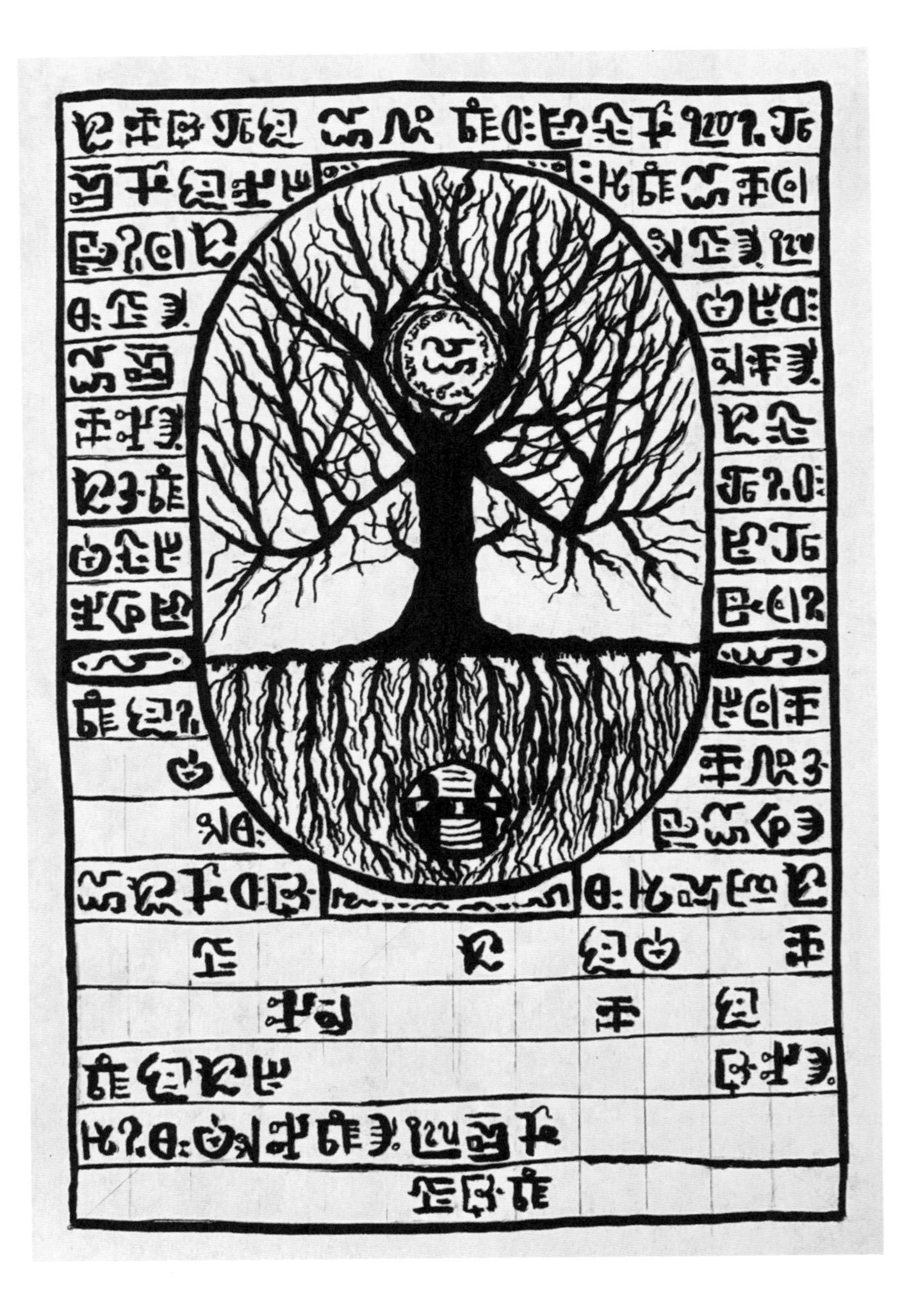

Art of the Grimoire

AN ILLUSTRATED HISTORY OF MAGIC BOOKS AND SPELLS

OWEN DAVIES

YALE UNIVERSITY PRESS

New Haven and London

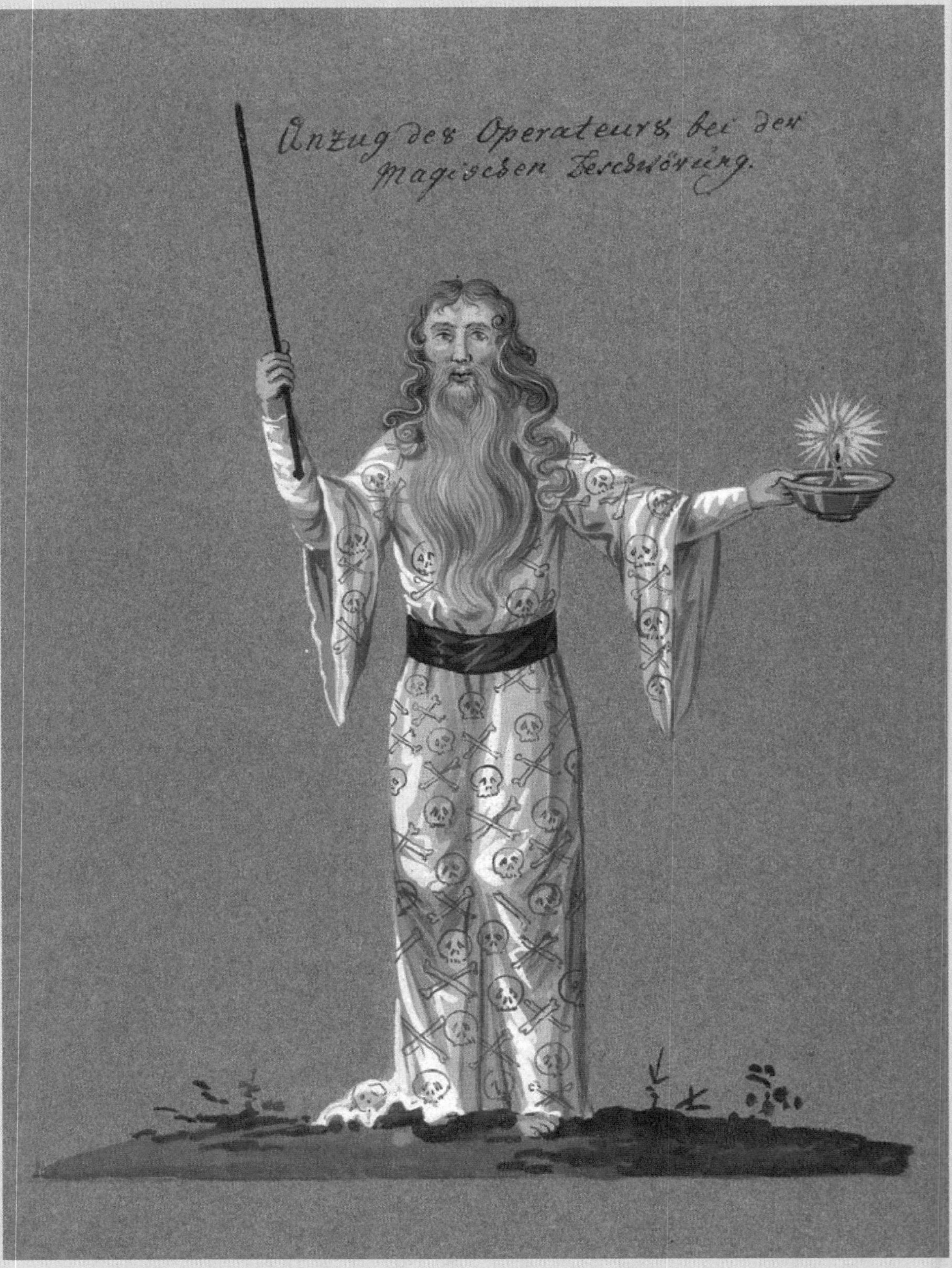
Anzug des Operateurs, bei der magischen Beschwörung.

CONTENTS

Page 2: "Grimoire #2," a drawing by American artist Jim Ingram, ink on silk. See page 227.

Page 4: Illuminated image of a magician from a rare, late 18th-century German-Latin grimoire, which contained beautifully painted watercolor drawings of demons and magical signs. The grimoire was entitled the *Compendium rarissimum totius Artis Magicae sistematisatae per celeberrimos Artis hujus Magistros* (*A rare Compendium of the entire Magical Art by the most famous Masters of this Art*).

Introduction

To understand the history of written magic is to understand the influence of the major religions of the world, the development of early science, the cultural power of print, the growth of literacy, the social impact of colonialism, and the expansion of esoteric cultures across the oceans. The contents of magical texts represent much more than magic itself, and their little-studied artistic qualities reveal unique traditions of illustration, design, and imagination. This book takes a global view, stretching from the ancient Middle East to ancient China, and from colonial Africa to the post-colonial Americas. It encompasses talismans for protection and simple spell books, as well as grimoires, which were manuals of magic that included a mixture of spirit conjurations and ritual instructions. Textual and pictorial forms of magic exist partly because of the need to create a physical record of magical knowledge and, in part, to perform magic. Yet, they also illustrate, quite literally, the human expression of fundamental desires, emotions, and fears in the form of demons, angels, spirits, gods, and abstruse symbols or abstract figures.

WHAT DOES MAGIC MEAN?

The word magic derives from the Greek *mageia*, which referred to the ceremonies and rituals performed by the *magoi*, or magi. These were thought to be priest–magicians from Chaldea, a Babylonian kingdom that existed at the head of the Persian Gulf from roughly the 9th to the 6th centuries BCE. From an ancient Greek perspective then, *mageia* was a foreign source of supernatural power, and we can see this idea of magic being the religion of other cultures over and over again. The spelling of magic with a "k" appears in 16th- and 17th-century English texts, such as *The History of Magick* (1657), which was a translation of a work by French librarian and physician Gabriel Naudé (1600–1653). At the time, the term "magick" did not signify anything different, it was simply a variant spelling. However, in the early 20th century, it was adopted by the notorious occultist Aleister Crowley (see Chapter 6, page 214), and given new cultural meaning. It distinguished supernatural or spiritual forms of magic from stage illusion tricks, but adding a sixth letter also gave it esoteric significance relating to occult hexagrams, the symbols of the power of six. Today, the term "magick" has been widely adopted by

LEFT Polished stone Neolithic axe head, inscribed in Greek (2nd to 3rd century CE). The text consists of a list of magical words, including Jewish names, which have been found on other gems and stones of the period. Known as "thunderstones," such Neolithic axes were thought to be the product of lightning strikes.

ritual magicians, and it is also used by the mass media when talking about modern forms of Western magic.

I have, in the past, described the attempt to define magic as a maddening task. Historians, sociologists, anthropologists, philosophers, folklorists, and psychologists have all tried to do so, over the last century and more. This is not the place to try and unpick the many debates. *Art of the Grimoire*'s focus on the interplay between text and image in the history of magic does, however, highlight the flawed, outdated, but still influential, anthropological notion that human cultures progress through three stages of development: the age of magic, followed by the age of religion, and finally the age of science—our modern world. The global history of magic shows, in fact, how magic, religion, and science have always been interlinked. Trying to set a defined boundary between religion and magic that covers all faiths and traditions is not possible. Some cultures, past and present, would see no distinction between the two, or would consider the religion of other cultures as magic and vice versa. As the many examples in this book show, religion can serve magic and magic can serve religion. As for science, the ensuing chapters demonstrate how advances in the technology of writing and communication enabled the recording, dissemination, and democratization of magical knowledge across time and space.

It is more useful, here, for the reader to understand some basic concepts of how magic was thought to function in cultures around the world over the millennia. One is the notion of sympathetic magic, which concerned the invisible relationship, or bond, between things either through contact, similarity, or ritual. The voodoo doll is a good example: stick sharp objects in the image of an enemy and they will feel the pain.

RIGHT Ethiopian Christian amulet scroll, written in *Ge'ez* sacred language on parchment (19th century). Such scrolls consist of incantation prayers against evil spirits and the health problems they cause, such as eye disease, and include the invocation of the secret names of God (*asmāt*). The imagery usually includes depictions of saints and angels.

Numerous anti-witch spells operated in this way. There was an invisible bond between witch and victim, and so rituals that involved the urine, body parts, or clothing of the bewitched would have a negative physical effect on the accused witch. Likewise, the act of drawing blood from a witch made the victim feel better. Then, there was natural magic, which included aspects of sympathetic magic, but also utilized the hidden properties of the natural world; in other words, the healing, harmful, or protective power inherent in gems, metals, plants, and animals, placed there by gods or spirits. Natural magic also encompassed alchemy, and the forces emanating from the stars and planets, and how they could be harnessed. Another branch of magic concerned the conjuration or adjuration of good and evil gods and spirits, either to expel, bind, control, or benefit from their knowledge and power. In addition, there were mechanical magical rituals and spells that did not require any form of external spiritual intervention to have an effect on the material environment. Some were designed to unlock personal human potential, while others were to influence the external world: do this action a certain way, say these words several times, draw this image, and x will happen.

BELOW Illustration from the *Clavis Inferni sive magia alba et nigra approbata Metratona* (*The key of hell with white and black magic proven by Metatron*). Created in ink and watercolor on vellum, it is written in Hebrew, Greek, Latin, and secret code. While dated 1717, it is most likely from the late 18th century. The image depicts four spirit kings and their animal forms, which could be conjured up by the magician.

ABOVE The *Great Pustaha*, a book of magic that belonged to a Batak (North Sumatra) magician-priest. Made from wood bark pages with an ornamental wood-carved cover, this piece is from the early 19th century, or possibly earlier. The earliest dated *Pustaha* derives from the archives of colonialization, with a Batak example donated to the British Library in 1764.

MAGIC AS ART AND ART AS MAGIC

In medieval Europe, the performance of ritual magic was sometimes considered an art within the framework of the seven "liberal arts" that underpinned education at the time, which were inspired by the intellectual ideas of the Greeks and Romans. These seven were: rhetoric, grammar, logic, music, geometry, arithmetic, and astronomy. Using magic in the quest for knowledge was defended by learned magicians as a pathway to acquiring the liberal arts, and it was even debated as to whether magic *was* one of the seven. After all, as we shall see, books of magic involved the construction of spoken and written languages, astronomical knowledge, the occult significance of numbers, and the power of geometric images. The same applies to ancient cultures in Asia, where the boundaries between art, science, and magic, as understood today, often evaporate when considering the means by which secret knowledge was thought obtainable—whether through prayer or the citing of mantras, or the act of writing, and sophisticated divinatory calculation.

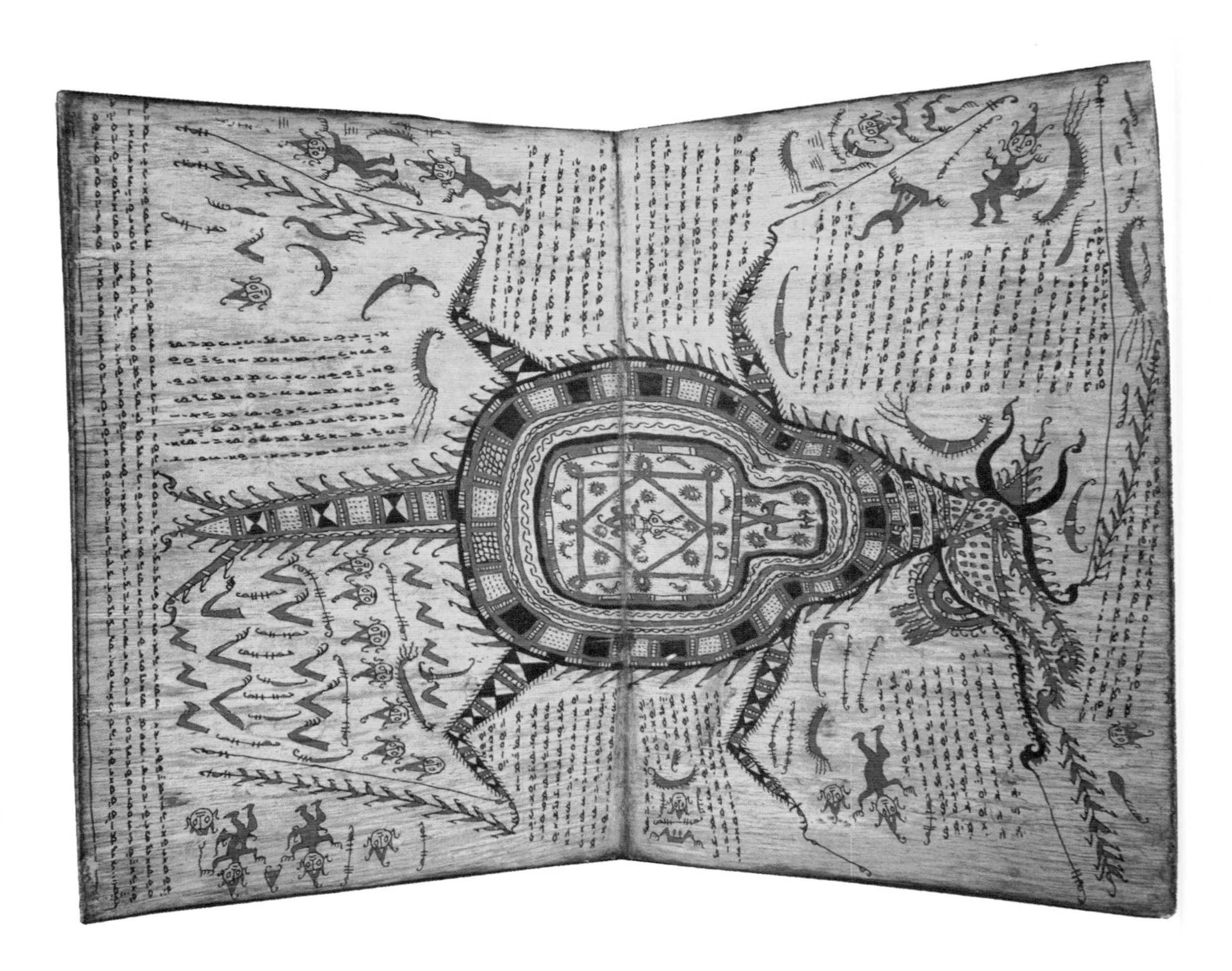

ABOVE Illustrations of mythical creatures from the *Great Pustaha*. The book, like others from the same regional manuscript tradition, consists of a series of glued, folding leaves that open like a concertina, rather than having separate pages and a spine. When fully unfolded, it is 56ft (17m) long. It contains a range of incantations and spells.

The use of graphic art *in* magic over the millennia was sometimes merely illustrative and decorative, but often meant so much more. Images of demons and spirits were a fundamental aspect of magic rituals, a means of contacting or controlling them, without which the accompanying written conjurations would not work. Religious images such as Christian crosses had protective functions, or served as meditative devices. Combinations of symbols, letters, and magical alphabets were used to create abstract art that also, collectively, held secret powers. Text and image were often literally entwined on the page, the potency of both inextricably linked. The grimoire, spell book, or talisman could, in itself, be a magical art object irrespective of its contents, serving, for instance, as the representation of the creative cultures of *others*, such as those magic texts from Africa and Asia appropriated by colonial antiquarians in the 19th century.

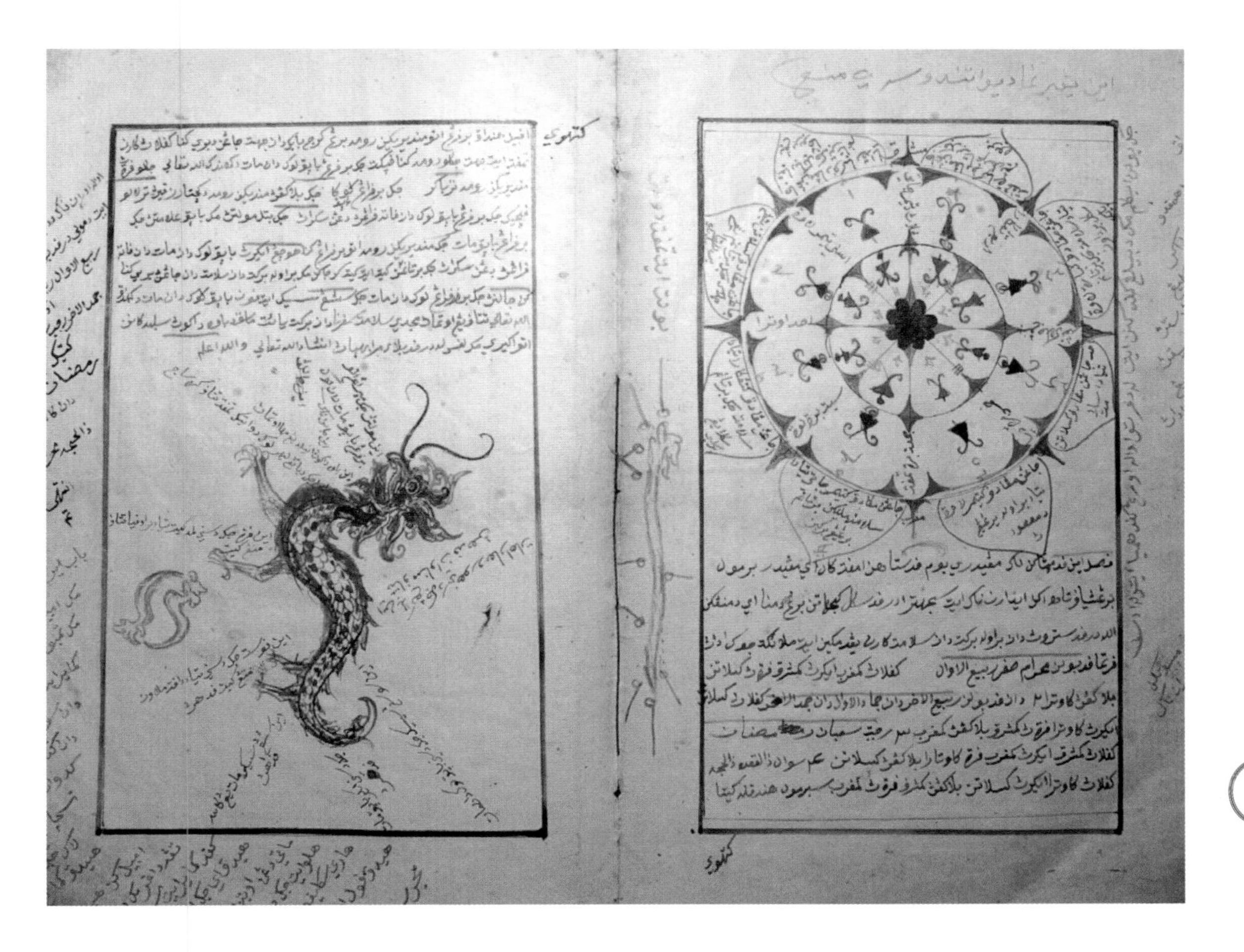

ABOVE Beautifully illustrated Malaysian Arabic magic and divination manuscript, copied by Zainal Abidin b. al-Marhum Tuan al-Haji Raja Kecik in 1894. It includes a *Rajamuka* wheel, which was used to convert words into a numerical identity for divination.

Finally, when we talk of the art in magic we must also think of those who created it—the priests, holy men, astrologers, doctors, intellectuals, and popular magicians, who carved, inscribed, and drew images and texts to impress clients with their arcane knowledge, pass on their wisdom, and, as ever, to make money. Some were mere copyists, yet others were true artists, whose imaginations were unleashed when putting magic on a writing surface. The technicians are also part of this creative tradition, whether the fabricators of early writing surfaces, printers and engravers in the age of print, or the games designers, film producers, and fiction writers of the contemporary world. The lineage of creativity in magic is long and it is as vibrant today as it ever was.

CHAPTER ONE | # Ancient Materials

The history of writing begins in Mesopotamia (a region that broadly encompasses present-day Iraq, eastern Syria, and southeastern Turkey) in around 3400–3200 BCE. The first evidence consists of pictograms or images of physical objects representative of ideas, and from this the cuneiform script developed, consisting of wedge-shaped marks impressed on clay tablets with a reed stylus. This first written form of communication reflected the language of the Sumerian civilization. Over the ensuing centuries the Sumerian script expanded to some 400 phonetic signs that expressed syllables. From around 2500 BCE, the function of cuneiform tablets developed from simple accountancy to records of religion, creative thought, and magic. Under the rule of the Akkadians, who succeeded the Sumerians as the great power in Mesopotamia, cuneiform was widely adopted across the Middle East. The ancient Egyptians invented their own form of pictographic hieroglyphic writing while cuneiform developed, with simple early examples carved on small ivory tablets acting as labels for grave goods rather than for administrative purposes. Then, a flowing cursive script known as hieratic developed as the language of everyday life in ancient Egypt, written using ink on ostraca (limestone flakes), wood, and later papyrus. Hieratic continued to be used for administrative, medical, and magical record keeping right down to the Graeco–Roman era (332 BCE–395 CE).

EARLY INSCRIPTIONS

The earliest written magic is found on Sumerian cuneiform clay tablets, thousands of which survive today in museum collections, providing an extraordinary insight into everyday ritual and belief at the time. From these, we discover that there were expert magicians or exorcists who were known in Akkadian as *āšipu* or *mašmaššu*, some of whom had libraries of clay tablets. Many of the surviving rituals and incantations concern protection from demons, ghosts, and those humans who were thought to perform harmful magic or witchcraft. Most contain instructions on how to create a range of protective ornaments and adornments that were widely used across society. There are guidelines, for instance, for making

LEFT Fragment of a cuneiform clay tablet containing a list of magical stones, Mesopotamia (mid- to late 1st millennium BCE). This is the sort of record that would have been kept by magician-physicians. Dozens of different amuletic and healing stones are listed here, but what they are in modern terms is not clear. Other similar lists of amulet stones survive from the period.

RIGHT Jewish Aramaic incantation bowl in terracotta and ink (c. 5th century CE). The inscription on this bowl is written in Jewish Babylonian, and it was commissioned from a magician-scribe by someone named Gia Bar Imma, for the purposes of an exorcism. The images depicted possibly represent the binding of two demons.

clay figurines using women's breast milk and white and black wool from a virgin female goat to protect the home. Other tablets instruct how to make amulets from magical herbs wrapped in sheepskin, which were then worn to deflect malevolent spirits. We also have examples of cuneiform writing on stone *stelæ*, or slabs, which were erected as public monuments to honor military victories, to mark territorial boundaries or sacred spaces, and to enshrine edicts and laws. Some contained ritual curses in the name of the gods.

While clay tablets provided a cheap writing surface as they could be produced in bulk using molds, they were rather heavy, and it has been estimated that if the contents of a 50-page modern, mass-produced paperback were transferred to clay tablets it would weigh over 110lb (50kg). Thin metal plates proved to be more portable for the creation of written amulets. Early examples of writing on gold sheets are rare but do exist for the Sumerian period. Considering the precious nature of gold, its use at this time as a writing surface was largely restricted to sacred purposes and kingship. The relative cheapness of lead, however, made it an ideal surface for inscribing more mundane information, and its soft, malleable nature meant it could be rolled up, folded, and easily carried. While lead was first extracted and used for making figurines some 8000 years ago, examples of flattened lead used for recording administrative information date back to the 8th century BCE in Anatolia (now in modern-day Turkey). Yet, the most well-known examples of lead writing are the Greek and Roman curse tablets, or *defixiones*. These thin sheets contained either simple curses against an enemy or requested such deities as Charon or Hecate to punish someone. One of the most common purposes was to exact vengeance on thieves. All but one of the curses found at the Roman baths dedicated to the goddess Sulis Minerva in Bath, were for this purpose. One example states, "Docimedis has lost two gloves and asks that the thief responsible should lose their minds [sic] and eyes in the goddess temple." Another, from a site dedicated to Mercury, begins, "Honoratus to the holy god Mercury. I complain to your divinity that I have lost two wheels and four cows and many small belongings from my house. I would ask the genius of your divinity that you do not allow health to the person who has done me wrong, nor allow him to lie or sit or drink or eat." These curses were not worn as amulets, but deposited at temples, thrown into holy springs, or buried in graves. Lead tablets were also used by Greeks and Romans for writing binding love spells or for aiding the troubled spirits of the untimely dead. It is also worth noting the tradition in late antiquity Egypt where curses were written in red ink or blood on large cow or camel bones.

Another tradition developed among the Phoenicians (1500–300 BCE), a great seafaring civilization that developed trading centers across the Mediterranean. In the era of their declining influence from around the 8th century BCE, archaeologists have found gold and

BELOW Byzantine bronze amulet in the Greek language (6th century). This piece demonstrates the melting pot of religious imagery on magical objects of the period. One side (left) depicts the figure of the Egyptian god Horus (see page 27), standing on two crocodiles with scorpions in his hands, surrounded by magical symbols. The other side depicts Christian scenes, including the Adoration of the Magi and Christ's healing miracles, and bears the inscription, "Lord, do not give strength to my enemies, for your right hand always protects me. Turn evils unto their heads, Lord our God; do not give them power."

silver sheets, or *lamellae,* inscribed with brief protective spells or prayers in the Phoenician Punic script. One example from Carthage, states simply, "Protect and guard Hilletsbaal, son of Arishatbaal." These sheets were rolled up and placed in ornate capsules to be kept about the person. Similar gold and silver *lamellae* were also popular across the Graeco–Roman world. Between the 3rd and 5th centuries CE a tradition of amulets also developed among the Mandaeans, an ethno-religious population with origins in Palestine. These usually consisted of densely engraved, rolled-up lead sheets, while a few examples made from gold and silver have also been found. A distinctive aspect of these Mandaic amulets is their lengthy first-person narratives about fighting off evil gods and spirits, as this extract from one such example shows:

> "Against Baqhus, father
> of curses, and against Karkum,
> father of sorcery,
> and against Tagya, mother
> of imprecations and Liliths
> who dwell on the gate of
> tombs and engraved
> mirrors; Gupa, the mother
> of wisdom will turn
> curse and imprecation
> and poverty against each one
> who envies me."

BELOW Roman curse on lead, Bologna, Italy (4th to 5th century CE). Written in Latin with Greek invocations, this piece contains a curse against a Roman senator named Fistus. Part of the curse reads: "Crush, kill Fistus the senator ... May Fistus dilute, languish, sink and may all his limbs dissolve." The inscribed image represents a gorgon-like demon with snakes around or attached to its head.

In China, writing developed independently around 1200 BCE, during the Shang Dynasty, with the earliest surviving examples of a fully expressive set of characters recorded from incisions on animal oracle bones used in divination rituals. But the earliest manufactured writing surface was made from bamboo, which had been split open, cut into strips, and dried. These strips were then bound together with silk. Once prepared, the scribe was ready to write in columns from top to bottom using a brush and black ink. The ink used was made from a base of finely ground burned ash and oil from pine trees, and animal collagen or gelatine was added as a glue. The mixture was dried into cakes for storage and transportation. Many of the early surviving bamboo writings are concerned with administrative issues, Confucian writings, and medical matters, but some are examples of *shushu*, or "calculations and arts," which covered divination, astrology, spells, and demons.

Two bamboo slips from the tomb of a Qin administrator dating to around 217 BCE contain an incantation prayer, or *dao*, against nightmares, which commands Qinqi the spirit master of night demons:

> "When a person has foul dreams, on wakening then unbind the hair, sit facing the northwest and chant this prayer: 'Heigh! I dare to declare you to Qinqi. So-and-so has had foul dreams. Flea back home to the place of Qinqi. Qinqi, drink heartily, eat heartily. Grant so-and-so great broadcloth. If not coins, then cloth. If not cocoons, then silkstuff.' Then it will stop."

We know from clay tablets that gemstones were thought to have magical properties in ancient Mesopotamia and Egypt—examples survive of images and symbols, often in the form of sacred animals or deities, engraved into or cut from precious stones, which were then worn as amulets and protective jewelry. The use of gemstones as a surface for writing magic, however, came into its own during the early centuries of the Common Era, produced for and worn by all sections of society and faiths. The innate power of the stone, along with a written spell, and magical or religious image was thought to be a potent combination. One of the more unusual genres of such amulets involved the repurposing of Neolithic polished stone axe heads in the Roman-era. These were engraved with words and images of power for protection, and were created particularly to prevent lightning strikes. Existing examples of these "thunderstones" show a mix of Hebrew, Egyptian, and Graeco–Roman influences, and the texts include magical palindromes, obscure words, and symbols of power.

RIGHT Vignette from an Egyptian magical scroll (664–525 BCE), ink on parchment written in hieratic script. The winged figure is a composite of several deity forms and appears to be concerned with the night. The associated text suggests that the image represents spells for the nocturnal protection of women and children from snakes and other dangers.

LEFT Engraved magnetite stone (1st to 4th century CE). This amulet is one of several surviving examples of magnetite gems depicting Eros, Greek god of love, embracing the mortal woman Psyche. As recorded in a magical papyrus of the time, the magnetic qualities of the stone held a sympathetic magical association with the power of amorous attraction.

EMERGENCE OF OTHER WRITING IMPLEMENTS

There is no doubt that the invention of papyrus was a revolution in the history of writing and the art of magic. It provided a light and receptive surface for writing and painting with inks. It also enabled the creation of book-length texts in the form of scrolls that could be tens of feet long. Whole histories, religious texts, and magical handbooks could now be produced in one continuous document. The fabrication of parchment is described elsewhere (see pages 66–67), so for the moment it is worth focusing on the technology and magic of ink. Derived from the Greek word *enckauston*, meaning burned or cooked, the earliest known inks were made from the soot of burned lamp oil, wood, or bones, that had been mixed with a little water and a binding agent, primarily gum Arabic from the acacia tree in the ancient Mediterranean world. One of the earliest uses of ink for magic is found in official hieratic execration texts written on bowls, statuettes, and dried clay. Dating from around 2600 to 1500 BCE, they were used to curse and banish enemies of the state and foreign rivals. The magic was enacted by smashing the inscribed objects and burying them at ritual sites.

Some papyrus manuscripts of the ancient Egyptian era used red ink made from iron-oxide pigment. Red ink was first used for writing headings on papyri texts, for instance, and also for indicating the beginning of a spell. The magico-religious world was also illuminated through the invention of other colored pigments for painting images on papyrus and stone,

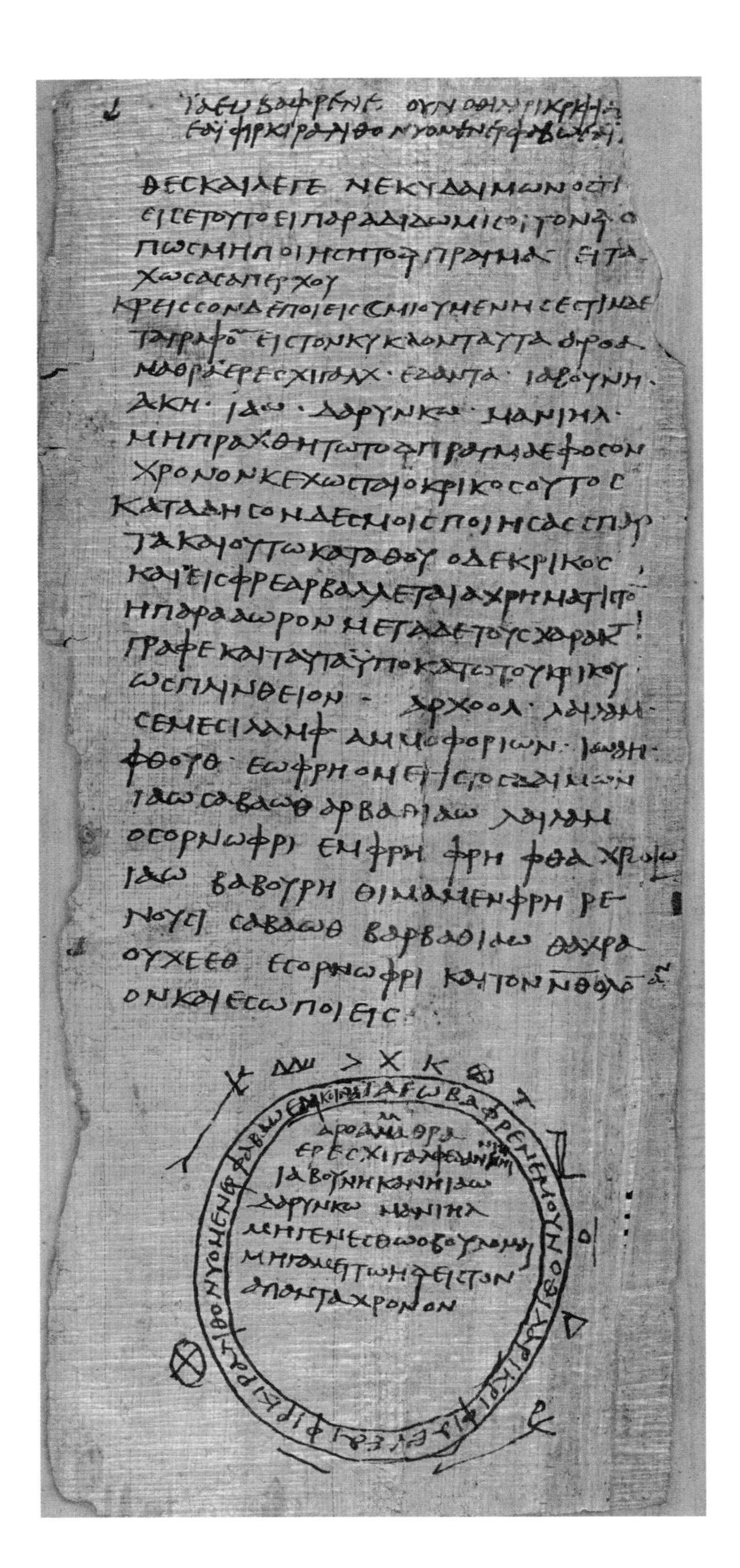

LEFT Page from a Greek "Handbook of Magic" (4th century CE), ink on papyrus. This offers instructions for requesting a spirit of the dead to prevent a marriage from taking place. The circle at the bottom of the page was to be inscribed on a lead *lamella* or on papyrus with myrrh ink from the outline of an iron ring. The writing in the center includes the words, "let whatever I wish not take place; let her, NN, not get married forever."

such as yellow ochre or Egyptian Blue, which was made from a heated compound that included copper and sand. Different inks could have specific properties for *enacting* magic as well as recording it. The corpus of the Greek Magical Papyri (PGM), a collection of documents that were created in Graeco–Roman Egypt, contains an instruction on how to write a protective amulet against demons on a piece of limewood specifically using vermilion ink. Vermilion, a red pigment made from the mineral cinnabar (mercury sulfide), had been used for wall decoration in both the ancient Mediterranean and Chinese worlds. Blood was another obvious source of potent red liquid, and Chinese Buddhist monks would prick their fingers to mix their blood with ink as a spiritual bond when making copies of *sutras* (the teachings of Buddha). There is less evidence of scribes or priests using human blood in the Mediterranean world, and magic handbooks mostly prescribed using animal blood rather than human. One such instruction from the PGM runs as follows:

> "Using blood from a quail, draw on a strip of linen the god Hermes, standing upright with the face of an ibis. Then with myrrh write also the name and say the formula: 'Come to me here quickly, you who have the power. I call upon you, the one appointed by the god of gods over the spirits to show this to me in dreams… Prophesy concerning this, concerning all things [about which] I inquire!'"

RIGHT Fragment of a birch bark scroll from the ancient region of Gandhāra (northern Afghanistan and Pakistan), written in black ink with a reed pen (c. 1st century CE). The numerous birch bark manuscripts found in Gandhāra (and in Gilgit, Kashmir) demonstrate the importance of the region to the spread of Buddhist religion and philosophy in all aspects, including folk magic.

While another example from a dream spell states, "Take red ochre [and blood] of a white dove and likewise of a crow, also sap of the mulberry, juice of single-stemmed wormwood, cinnabar and rainwater and write with it and with black writing ink and recite the formula."

The *performance* of magic in late antiquity, as well as the recording of magico-religious rituals, is also demonstrated by the survival of over 2000 incantation bowls from the region that is modern-day Iraq. These domestic earthenware bowls date from the 3rd to 7th century CE and have been found ritually buried upside down in domestic buildings and cemeteries. The spells were written inside the bowl in concentric circles using ink. Most of the surviving examples are written in Aramaic with quotes from scripture, showing the influence of Jewish and Christian religion, although some Mandaean, Persian, and Arabic examples also exist. Many bowls include images depicting demonic figures or demons in animal form. One bowl incantation refers to "all the evil" in "the likeness of vermin and reptile, in the likeness of beast and bird."

The texts in the bowls show that they had an apotropaic function, meaning they were created to protect the home from demons and unwanted spirits. Although they are sometimes referred to as "devil-trap bowls," the incantations actually make no reference of their use to confine devils. Why would you want to trap one in your home? In fact, the aim was to expel them. The following example concerns Lilith, a Judaic and Mesopotamian she-demon:

> "This is the amulet against the Lilith which is lurking in the house of this Epra the son of Šaborduk and this Bahmanduk the daughter of Sama. I adjure you, every species of Lilith, in the name of your offspring which demons and liliths bore for the children of fire who went astray."

It's important to note also that across the ancient world, magic was written on a range of untreated organic surfaces that rarely survive. A remarkable trove of early Buddhist manuscripts written on birch bark dating to the 5th, 6th, and 7th centuries CE were discovered in the 1930s, in the ruins of a religious establishment in Gilgit, Kashmir. They include numerous spells and apotropaic instructions, which the monks were presumably providing as a service to the community. We know that in late antiquity some protective amulets consisted of text on eggshells, leaves, and fruit. Indeed, we have instructions on how to ingest magical or religious words written on edible objects: one Jewish magic spell for improving the memory or "Opening of the heart" was to "write Psalm 119:97 on an apple or an ethrog and eat it;" another concerning impotency instructed the reader to write it "on three leaves of an elm tree and wipe it away with water from a jar. And give the bridegroom and bride [the water] to drink." Elsewhere, the magic of erasure—drinking water that has been poured over a freshly written piece of magic or holy text—was practiced in Islamic, Christian, and Jewish popular religion.

LEFT Mandean incantation against evil spirits on lead, Iraq (5th to 7th century CE). This incantation is similar to those found on Mandean incantation bowls. It includes a long list of demons, some associated with specific locations, and the binding of curses, including "the three hundred sixty curses of darkness which exist in the firmament, in the four corners of the world."

ABOVE Mesopotamian cuneiform clay tablet (mid- to late 1st millennium BCE). The text on this piece relates to a person tormented by a ghost, and calls upon the Sumerian gods Gula and Asalluhi to help cure the patient. Gula was an important healing goddess, sometimes represented as a midwife, while Asalluhi was the god of exorcism. The pair were frequently invoked together in magical-medical texts of the period.

RIGHT Small magical stela to Shed, dedicated by Nesamenemopet (c. 750–664 BCE). Shed (often called the "Savior") was a variant of the falcon-headed Egyptian god Horus, but he was less of an official deity of worship and more of a god of the common people. He was usually depicted as a child or young man. Numerous stelae dedicated to him have been found, in particular at the ancient settlement of Deir el-Medina, Egypt, and they served to protect the community.

LEFT Magical stela (c. 360–343 BCE). This stela was commissioned by a priest named Esatum for a temple to sacred bulls in the ancient city of Heliopolis, now a suburb of Cairo, although it was later removed to Alexandria. This piece is part of a genre of stelae dedicated to Horus, the falcon-headed god, for protection from poisonous animals. Horus (the central figure on the stela holding scorpions) was the son of Isis and Osiris, who was stung by a scorpion or bitten by a snake as a child, according to myth. Thus the stela records the following spell: "Don't fear, don't fear, my son Horus! I will be around you as your protection and drive all evil from you and any man who is suffering as well." People drank water poured over the stela as an antidote against poison.

ABOVE Boundary stone from Nippur, Iraq (1146–1123 BCE). It was discovered in 1896 on the northwest side of the ziggurat (religious tower) in Nippur, within the temple precinct. As well as an historic account of the ownership of the land and hymn to the great Sumerian god Enlil, the stone is inscribed with written curses that invoked the gods to punish anyone who tried to take possession of the land on which it stood, or attempted to destroy or remove it.

Papyrus

Papyrus is made from a type of sedge plant of the same name (*Cyperus papyrus*), which grew abundantly along the Nile Delta in ancient times. It was called *aaru* by the Egyptians, which also described the "Field of Reeds"—a paradise presided over by the great god Osiris. In more mundane terms, it was a plant with multiple valuable uses. It was burned for fuel, the stems were used to make boats, baskets, and mats, the roots were carved to make utensils, and the pith was eaten. The stem was first processed to create a writing surface in the early part of the third millennium BCE. The exact production process in early antiquity remains something of a mystery and modern attempts to recreate it do not always match the archaeological evidence. The first surviving detailed description comes much later, from the Roman naturalist Pliny the Elder (d. 79 CE), although it is unlikely he ever saw it being made at first hand.

The many surviving magical papyri contain numerous accounts of the magical properties of plants, animals, and gems, but it seems the papyrus

BELOW The *Book of the Dead*, papyrus of Ani (c. 1250 BCE). The image shows the ancient Egyptian goddess Hathor descending into the papyrus marshes in the form of a cow. Hathor was a state-worshipped solar goddess, who held power over love, fertility, and pleasure. The figure to the left of Hathor is the household hippopotamus goddess Taweret, associated with fertility, who carries water lily offerings to provide strength.

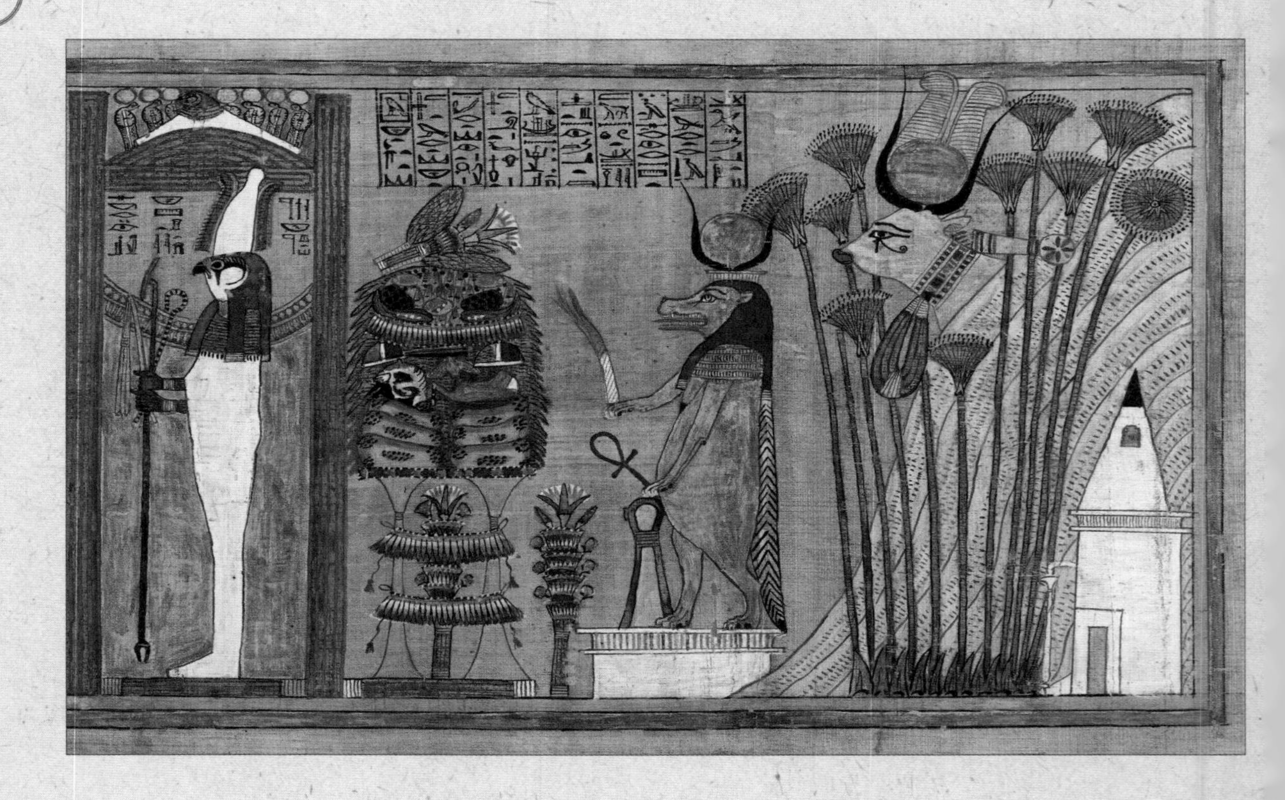

ABOVE The first step in making papyrus was to cut long, thin strips from the pith of the green stalks of the plant. Modern experiments have shown this to require a skilled hand. The strips may have been soaked, before being laid across each other, as shown above, and then pressed and dried. The exact method is unrecorded. The natural gums in the pith acted as a bonding agent.

plant was rarely included among them. Despite the religious significance of the plant, it seems the importance of papyrus was primarily as a record of knowledge rather than as an object with any intrinsic spiritual or magical power. However, it was thought to have important medical properties and what was described as "unwritten papyrus," particularly when burned to ash, was used widely in a variety of medical remedies as reported by the Egyptians, Greeks, and Romans.

As a writing surface, papyrus does not survive well in the cooler, damper climates of Europe and so never became a medium of record in the region. It was prone to molds and probably disintegrated within decades. It was only in the second half of the 19th century that the extensive use of papyrus for writing in the ancient world first came to be understood, and it was not until the 1880s that collectors, museums, and scholars really began to grasp the cultural and magical wealth recorded on humble papyrus.

LEFT Aramaic incantation bowl from Nippur, Iraq (3rd to 7th century CE). The Aramaic text consists of two charms wherein two magicians (pictured at the bottom of the bowl) each invoke their powers on the other's behalf to protect their homes. One begins, "Again I come, I Pâbak bar Kûfithâi, in my own might, on my person polished armor of iron, my head of iron, my figure of pure fire." Then he commands, "I will lay a spell upon you, the spell of the Sea and the spell of the monster Leviathan. (I say) that if at all you sin against Abûnâ b. G., and against his wife and his sons, I will bend the bow against you and stretch the bow-string at you."

ABOVE Chinese oracle bone (late Shang dynasty, c. 1300–1050 BCE). Most oracle bones and oracle tortoise shells have been discovered around the ancient city of Yin, the capital of the Shang dynasty (1600–1046 BCE), now modern-day Anyang city in Henan province. In the 19th century, locals working in the fields around Anyang found strangely incised animal bones and shells, which they called dragon bones. These became highly prized in local traditional medicine, and were ground up and put in herbal potions. Apparently, one day in around 1899, the antiquarian Wang Yirong (1845–1900) was drinking a herbal soup for his malaria when he found a bit of bone with strange incisions floating in it. He subsequently purchased some whole bones from the local apothecary, and realized the importance of his discovery: the incisions were the earliest known form of the Chinese language.

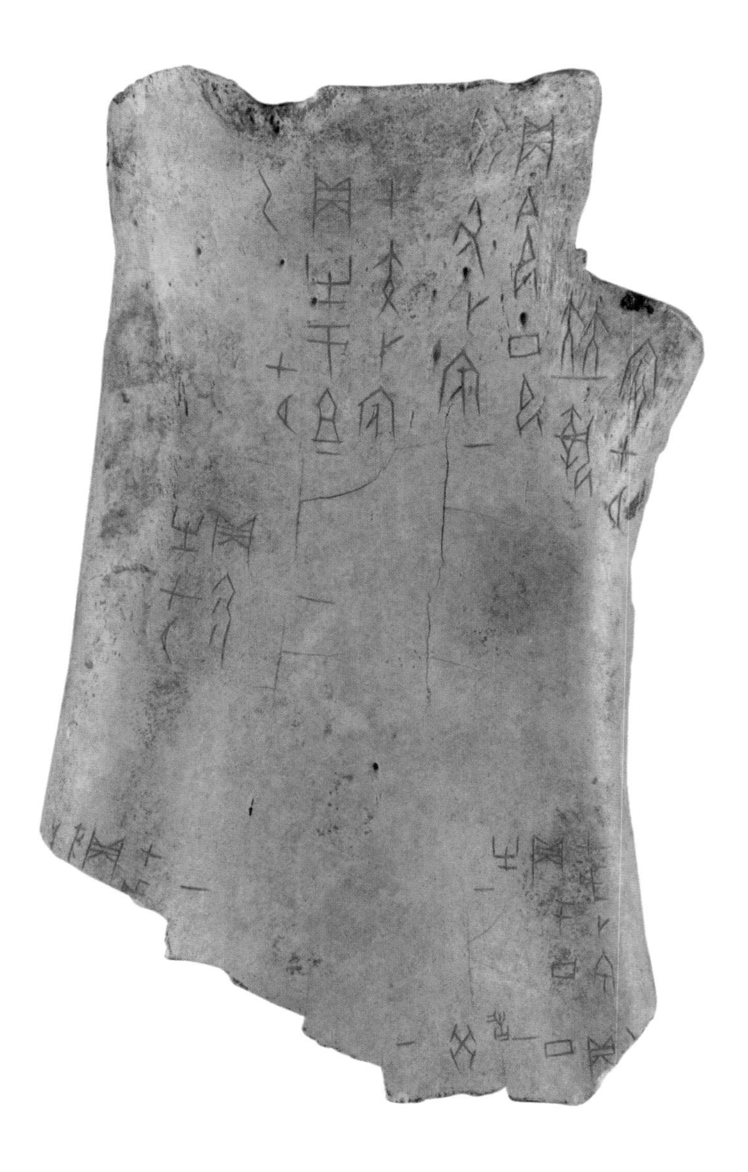

ABOVE Chinese oracle bone (Shang dynasty, 13th century BCE). Most oracle bones have been found in fragments (here, a piece of ox shoulder blade), but the few that have survived whole or have been pieced together, reveal a highly institutionalized culture of divination from the royal court to villagers. The bones were carefully prepared, before priest-diviners wrote questions on them on behalf of their clients relating to war, weather, crops, and politics. These questions were addressed to the gods, and the bones were then scorched by an extreme heat source, with the resulting cracks interpreted. Sometimes, the outcomes were recorded on the bones. The descriptions are laconic, though, such as a diviner named Zi who tested the proposition, "Sacrifice a small set of penned sheep to the Dragon Mother [Star]." Another, which included the answer, records: "If we sacrifice to the rising of the Dragon [Star] in the fields at Fan, there will be rain; auspicious."

ABOVE Aramaic skull incantation (5th to 7th century CE). Several such skulls have been found, though their provenances are sketchy. This example from the Vorderasiatisches Museum, Berlin, is the best preserved. The skull of a 30- to 50-year-old, the composition and contents of the inscription are similar to those on incantation bowls of the period. Although the intended outcome of the spell is not entirely clear on this example, it refers to a demon named Bartytba, son of the sister of Libat/Dlibat (a goddess of love) and the great grandson of Lilith (the she-demon). It also refers to the Judaic angels Nuriel, Hashtiel, Shibiel, and Gabriel.

Bamboo Slip Discoveries in China

In collections today, there are around 300,000 Chinese bamboo slip manuscripts dating from between the 4th century BCE to the 10th century CE. They have also been found in Korea and Japan. Their survival is dependent on being sealed in dry conditions in ancient times. The silk binding threads that held the bamboo strips together have usually long rotted away, leaving an interpretive puzzle for archaeologists.

The survival, or not, of such ancient bamboo texts was not only a matter of archaeological processes, though. The first emperor of a united China, Qin Shi Huang (c. 259–210 BCE), who put an end to the Warring States period that had begun in the 5th century BCE, ordered the destruction of all history writings to ensure his reign could not be compared with those of previous emperors. It was clearly not entirely successful, and, thankfully, his edict did not extend to books of astrology, medicine, and divination. One of the most extensive collections that survived the emperor's edict is now known as the Tsinghua

BELOW A book composed of bound bamboo slips, or strips, featuring a weapons directory, from North China (c. 93 CE, Han dynasty). The slips were inscribed with brush and ink, and then joined together with either silk or, as shown here, with hemp string. Once completed, the bound slips were rolled into a scroll, which was called *Jian Ce*, or *Jian Du*, for storage or carrying.

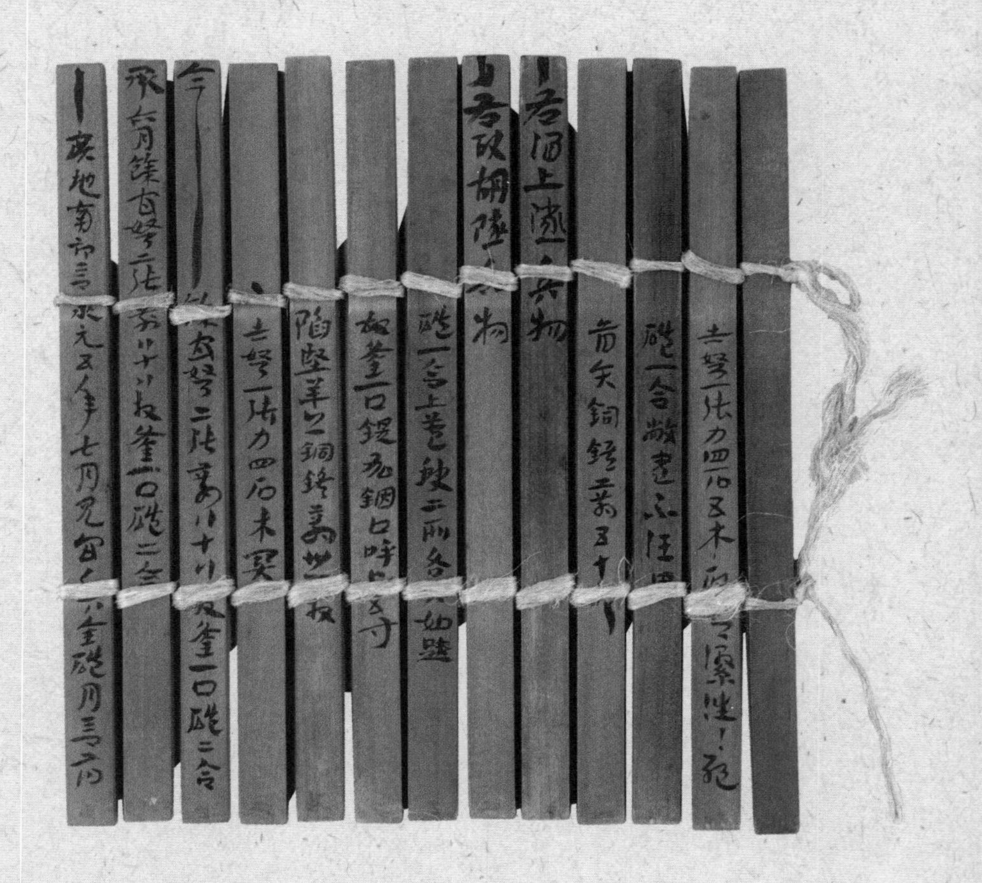

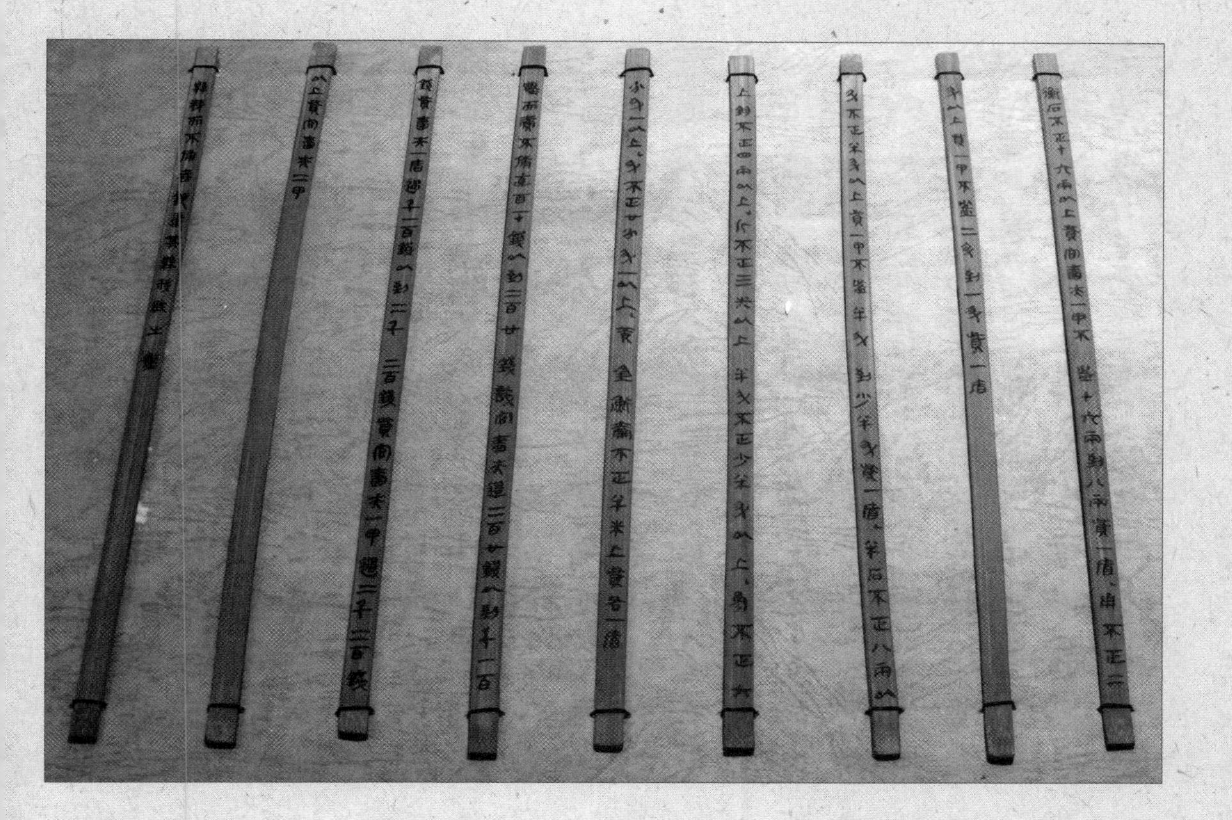

ABOVE Examples of unbound bamboo slips from tombs of the Qin Dynasty at Shuihudi (c. 217 BCE). The slips dictated the development of the form of Chinese writing, which was written in columns from top to bottom and from right to left. The author corrected writing mistakes by scraping them away with a small knife, which became a symbol of scribal power.

Bamboo Slips, consisting of nearly 2400 items that were written down around 305 BCE. Like numerous other collections, these slips were uncovered by illegal tomb robbers, before being acquired by Tsinghua University in 2008. One of the most important of the Tsinghua bamboo documents consists of what is probably the first decimal multiplication table in the world, but the corpus also contains instructions on geomancy, as well as advice on dealing with spirits and demons.

A common occult topic found in Chinese bamboo collections concerns the identification of auspicious days and the avoidance of unlucky days. The Zhoujiatai collection, excavated from the tomb of a minor official in Hubei province in central China in 1993, contains an advice manual on such matters consisting of 73 slips. One example, which dates to around 209 BCE, has these guidelines:

"When travel is urgent and you cannot wait for a good day: when traveling east overcome Wood; when traveling south overcome Fire; when traveling west overcome Metal; when traveling north overcome Water. It is all right to not wait for a good day."

Another bamboo slip collection from the same province, excavated in 2000, explained the meaning of this text. It told that iron conquered wood, carrying a vessel of water overcame fire, charcoal conquered metal, and keeping earth in a cloth countered water.

ABOVE Pella curse scroll (375–350 BCE). Made of lead, and named after the ancient Macedon capital where it was excavated from a grave, the Pella scroll is inscribed in a Doric Greek (also known as West Greek) script. It was written by, or for, a woman who desired to marry her lover named Dionysophon: "for I wish him to take no other woman than me, and that [I] grow old with Dionysophon, and no one else." Yet, this love spell contains a curse, directed at her rival, a woman named Thetima, whom she desired to perish miserably.

ABOVE Roman curse on lead, found in 1927, in the arena of the amphitheater at Caerleon, Wales. The inscription reads: "Lady Nemesis, I give thee a cloak and a pair of boots; let him who wore them not redeem them except with his life and blood." Nemesis was the goddess of justice and was consequently honored, in particular, by soldiers, priests, and officials. Small shrines to Nemesis have been found at amphitheaters elsewhere in the Roman world.

LEFT Onyx amulet of unknown provenance, although likely Eastern Mediterranean (5th to 6th century CE). The image on the front depicts the Akedah, or Binding, of Isaac (as related in Genesis 22). It shows Abraham, knife in hand, with his son Isaac bound and ready for sacrifice on the altar on the orders of God. The hand descending from Heaven is that of an angel who intercedes at God's behest, and replaces Isaac with a ram. On the reverse side of the amulet are four lines of indecipherable pseudo-Aramaic. Depictions of the Akedah appear on other amulets and seals from late antiquity.

ABOVE Green jasper amulet (1st to 5th century CE). The two sides represent the rich fusion of religious and magical cultures in the Eastern Mediterranean with Roman, Greek, Egyptian, Jewish, and other Semitic influences. Around the edge is the *ouroboros*—a symbol of a serpent swallowing its tail that dates back to ancient Egypt. The eagle-headed god and cock-headed god with snake legs are also both Egyptian in origin, although the latter wears Roman dress. The letters are Greek, and the Judaic angels Michael, Raphael, Gabriel, and Ouriel are listed above the shield of the cock-headed god.

Coptic Magic

One amazing collection of papyrus magical handbooks was produced in Coptic Egypt. Coptic was a group of Egyptian dialects spoken by common people in Egypt between the 4th and 8th centuries CE that used the Greek alphabet in its written form. While it was eventually supplanted by Arabic, it remained the language of the Coptic Orthodox Church in Egypt. The Coptic magic scrolls reflect, therefore, the influence of early Christianity with the appeal to saints, the invocation of holy names, and references to biblical apocrypha sometimes blurring the boundary between prayer and incantation. There are clear examples where pre-Christian Egyptian magical formulations were revised to make Jesus the omnipotent power, while the rest of the incantation text remained largely the same. The following Coptic text is a good example, where God is beseeched to send Christ to dispel "every unclean spirit of the dirty aggressor, from 100 years downward and for 21 miles round, be it a male demon, be it a female demon, be it a male poison, be it a female poison, be it a vain, ill-bred, dirty demon!"

A major research project, "The Coptic Magical Papyri: Vernacular Religion in Late Roman and Early Islamic Egypt" (2021–2023), led by scholars at the Julius Maximilian University of Würzburg, has calculated that some 600 Coptic magical texts on various writing surfaces survive today, in collections dating from the 3rd to the 12th century. The two most common purposes for magic in the corpus were concerned with healing and protection, followed by cursing and love magic.

Excavations at the ancient oasis town of Kellis, some 200 miles inland from the Nile, have revealed several fascinating Coptic magical texts. One concerns an exchange of papyrus letters between an inhabitant called Pshai and a man

RIGHT Coptic magical figures, drawn in iron gall ink on papyrus (8th to 9th century CE). Much of the surviving corpus of Coptic magic consists of such fragments, which are not always easy to interpret. Crude depictions of angels and demons are quite common, although sometimes it is only through the text that the meaning or identity of the images become clear.

LEFT Papyrus codex (see also page 50), with leather cover and parchment decoration, probably from Thebes (6th century CE). Over 16 pages, the Christian author charts a careful course through what is magical and what is liturgical. It includes an apocryphal exchange of letters between Jesus and King Abgar V of Osroene (Upper Mesopotamia), to be copied for protection.

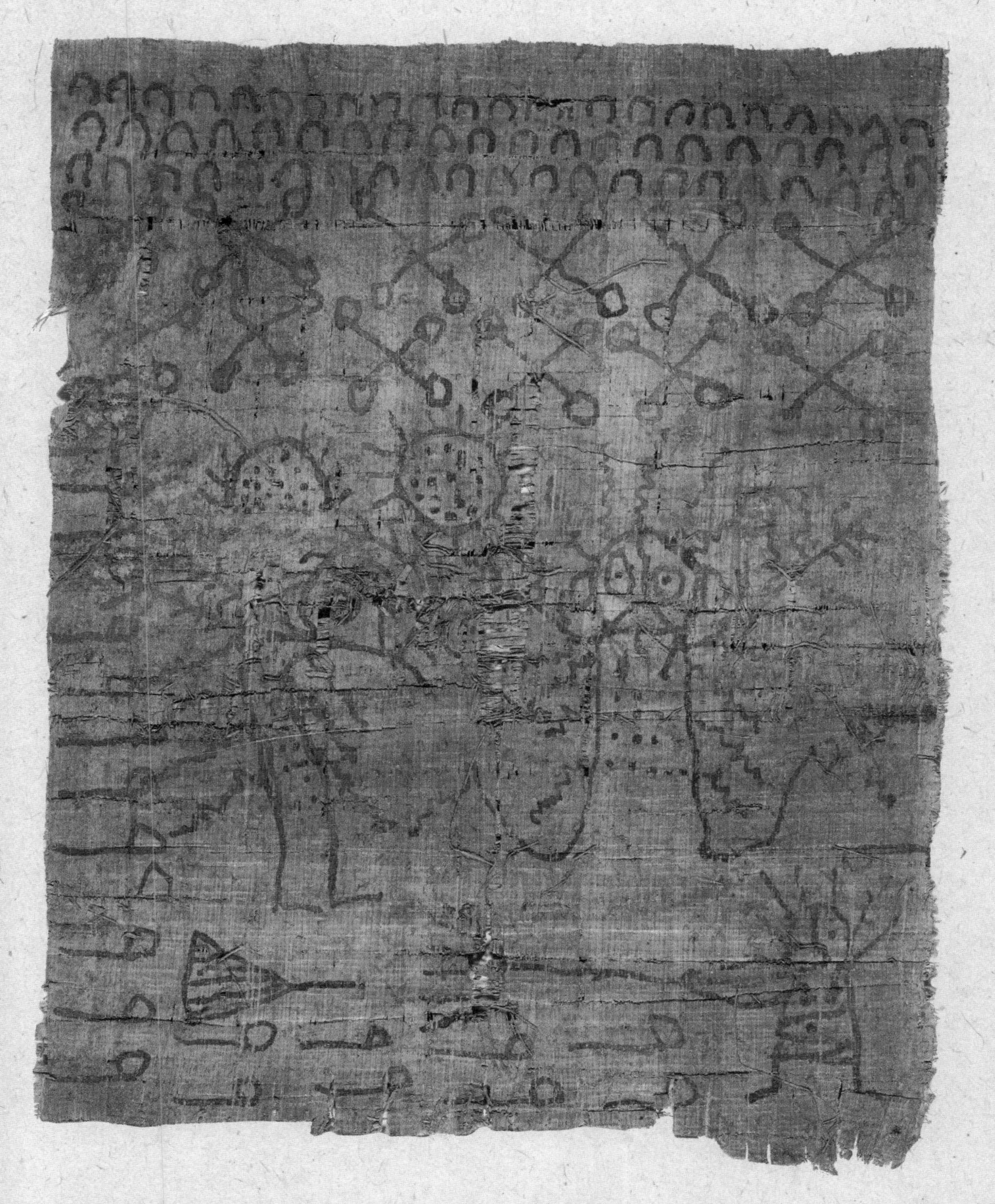

named Ouales, who was evidently thought to have good magical knowledge. Pshai urgently requested a separation spell to break the love of another couple, which Ouales supplied. This instructed Pshai to burn mustard seed, which, through sympathetic magic, would burn and blacken the hearts of the lovers. The exchange shows how papyrus was not only a record of magic but also an influential communication tool for obtaining magical instruction.

ABOVE Book of the Dead of Neferrenpet (1279–1213 BCE). In ink on papyrus, this is a chapter from the *Book of the Dead*, also known as the *Book of Coming Forth by Day*, of an ancient Egyptian relief sculptor named Neferrenpet. Such books, which were written by priests, contained a series of spells illustrated with supporting vignettes that were to aid a dead person through their journey in the afterlife. There is no strict uniformity in the contents of these papyrus scrolls, with some evidently commissioned by and tailored to the needs of high-ranking individuals. The right-hand vignette shows a figure kneeling and worshipping Osiris, the god of death and the afterlife, along with two other deities before a gate.

ABOVE Vignettes from the *Book of the Dead of Reri* (305–30 BCE), written in hieratic script on papyrus (see larger section overleaf). The top image depicts a snake being driven away, along with the associated text, which is found in other Books of the Dead and on tomb walls. It reads: "Oh Rerek snake, take yourself off, for Geb protects me, get up, you have eaten a mouse which Ra detests, and you have chewed the bones of a putrid cat!" Geb was the grandson of Ra, and god of earth. The second image concerns a spell for the deceased to use against crocodiles, and begins, "Get back, you dangerous one! Do not come against me, do not follow my magic," and ends with, "no crocodile which lives by magic shall take [my magic] away!" In ancient Egyptian belief, dangerous animals such as snakes and crocodiles were just as much a threat in the spiritual world as they were in earthly life.

OVERLEAF Section from the *Book of the Dead of Reri*. Reri was a priest at a temple in Thebes, who bore the title "Second god's servant of Amun." He was in charge of the administration of the temple and its workers, as well as managing offerings to Amun, the patron deity of Thebes.

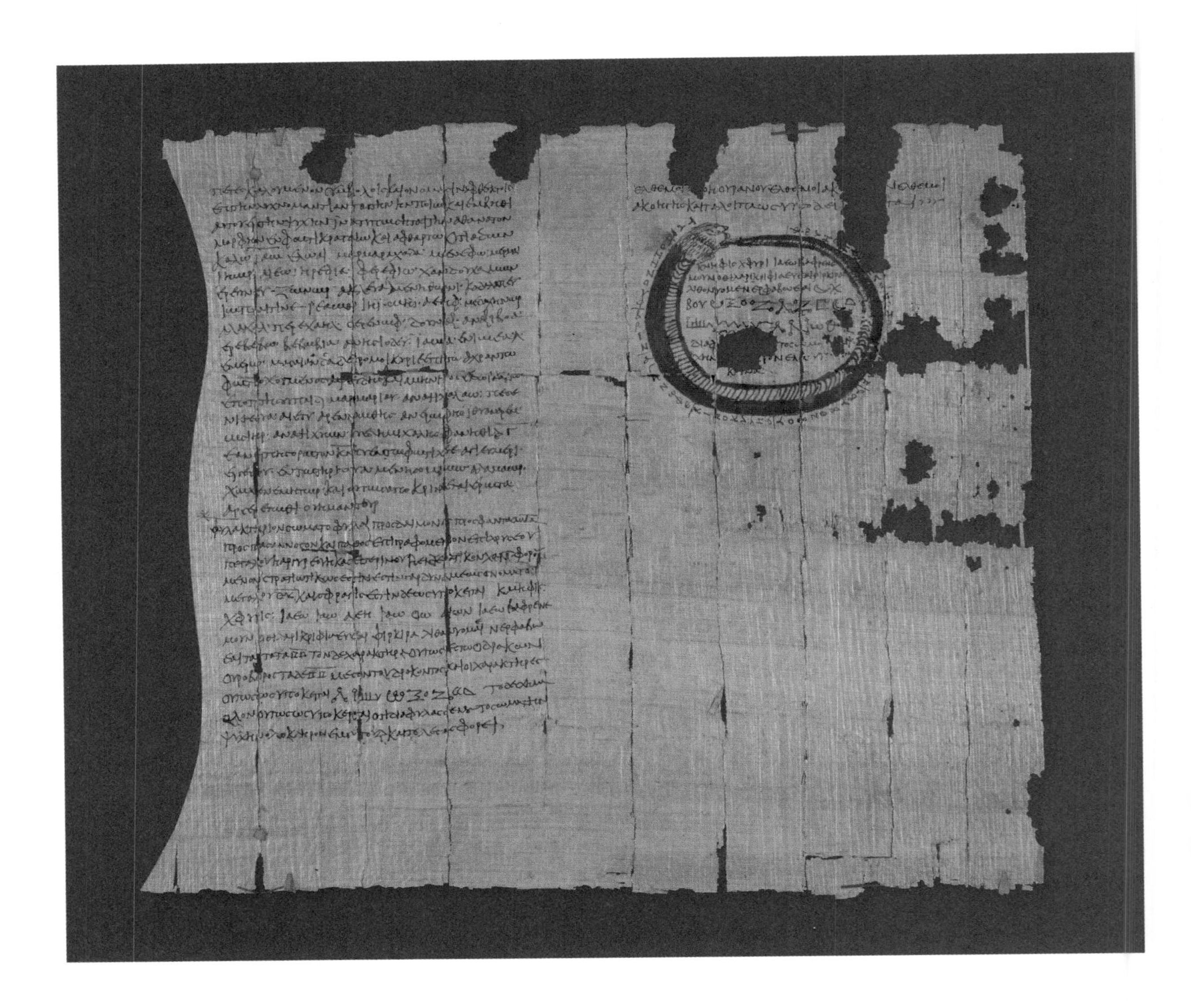

ABOVE Portion of a Greek papyrus magical handbook, possibly from Thebes (3rd century CE). The scroll contains a wide range of spells, including to keep insects out of the house, induce insomnia by means of a bat, win at dice, stop demons and apparitions, silence others, and provoke love. It also includes a cure for bad breath after eating garlic. The charm depicted here, includes an *ouroboros* (serpent swallowing its tail), which was drawn as a protective amulet to be worn against harmful spirits and illness.

TOP RIGHT Papyrus love attraction spell written in Greek (4th century CE). The main figure here is the god Bes, the Egyptian protector of households, who is shown sticking out his tongue. The second figure holds a knife and a severed head. The text entreats, "I adjure you by the twelve elements of heaven and the twenty-four elements of the world, that you attract Herakles whom [Ta]epis bore, to me, to Allous, whom Alexandria bore, immediately, immediately; quickly, quickly."

BOTTOM RIGHT Erotic magical papyrus in Greek language (4th century CE). This text includes a range of spells, some with images to be drawn as part of the magic, mostly relating to matters of love, but also charms to restrain anger and for opening a door. The image on the left is a cock-headed demon found in other love spells of the period. The image and symbols on the left were to be inscribed on a silver *lamella* and worn under one's clothing to win favor and to be victorious in legal cases.

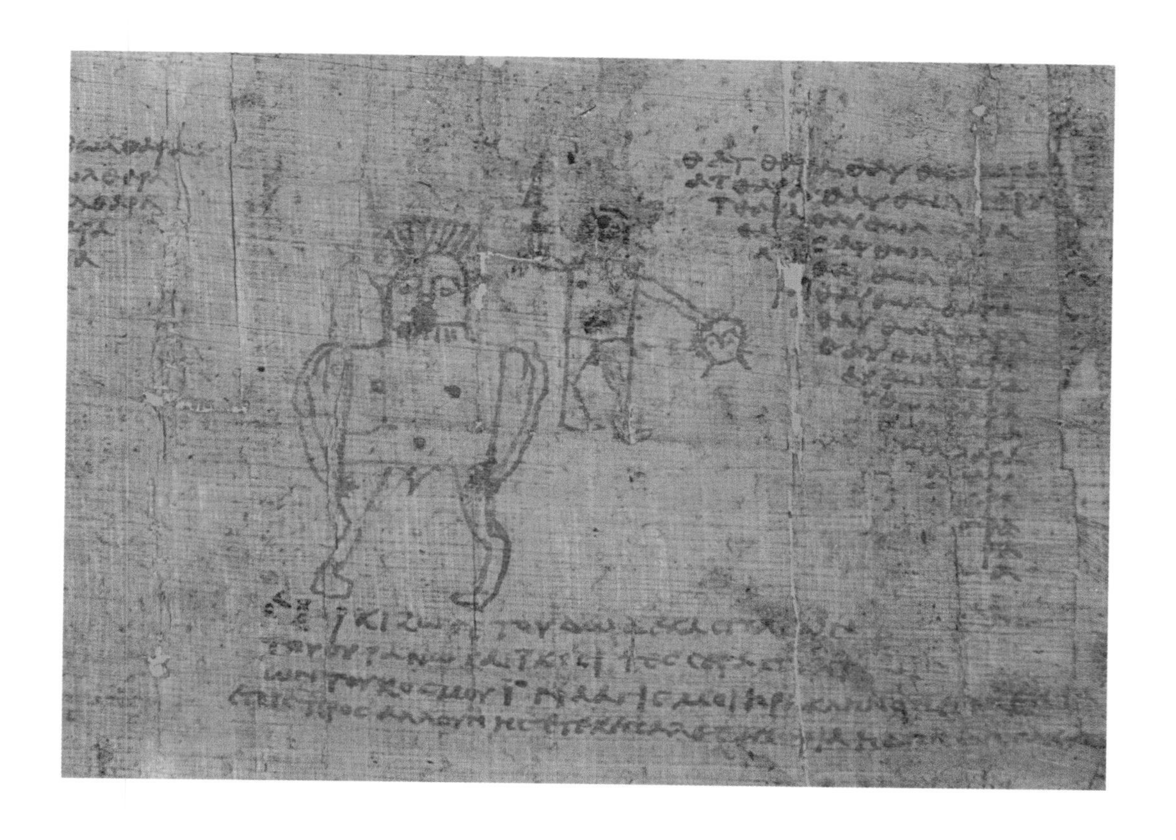

LEFT Papyrus charm written in Greek (2nd century CE). The text here reads: "'PHNOUNEBEE (two times), give me your strength, IO ABRASAX(?), give me your strength, for I am ABRASAX.' Say it seven times while holding your two thumbs." Abrasax (or Abraxas) was a word of magical power at the time, and was also the name of a demon deity of protection, frequently called upon in Greek papyrus spells.

RIGHT Coptic parchment (10th to 11th century CE). This parchment sheet, which illustrates the growing shift away from the use of papyrus in Egypt in this period, contains two recipes. One is a curse, and the other a love spell to attract a woman, which involves an offering that includes burned cotton and pepper.

ϯⲥⲟⲡⲥ ⲁⲩⲱ ϯⲡⲁⲣⲁ
ⲗⲓⲛ ⲙⲙⲁⲕ: ⲙⲡⲟⲟ
ⲃⲁⲣⲟⲩⲭ ⲁⲡⲛⲟϭ
ⲛⲁⲧⲱⲥ ⲭⲉⲕⲟⲁⲥ
ⲧⲉⲩⲛⲟⲩ ⲛ̄ϣⲁⲓⲧⲱ
ⲙⲁⲕ: ϩⲓ ⲡⲣⲟⲉⲡⲏⲓ
ⲁⲩⲱ ⲧⲁⲙⲟⲩⲣ ⲙⲁⲕ:
ϩⲓⲇⲉⲛ ⲡⲉϭⲛⲁϩ

ⲗⲟⲕⲉϯⲛⲟⲩⲕⲁⲣⲓⲱⲃⲟ ⲉⲧⲧⲁⲡⲣⲟ ⲉϩⲟⲩⲛ: ⲁⲩⲱ ⲡⲉⲛⲧⲁⲉⲃⲁ
ⲣⲱⲙⲉⲛⲓⲙ: ⲛⲉⲕⲟⲩⲓ: ⲙⲛ ⲛⲉⲛⲟϭ: ⲙⲛ ⲙⲏⲧⲉ ⲛ̄ⲡⲟ
ⲛ̄ⲛⲟⲩⲭⲁⲗⲓⲛⲟⲥ ⲙ̄ⲛⲟⲩϣⲧⲁⲙⲉ: ⲉⲧⲧⲁⲡⲣⲟ
ⲟⲩⲛ: ⲙⲛⲟⲩϣⲱϣ ⲙⲛⲟⲩⲕⲱⲛⲧⲏⲩ: ϩⲛ ⲧⲉ
ⲟⲩ: ⲡⲁⲧⲉⲕⲉⲑⲓⲁⲛⲉⲓ: ⲡⲉⲥⲙⲟ[...]ⲑ ⲛ̄ⲧⲁⲕ
ⲧⲁⲙⲉⲣⲱϣ: ⲛⲉⲛ ⲙⲟⲩⲓ: ϩⲓⲑⲏⲛ: ⲇⲁⲛⲓⲏⲗ
ⲡⲣⲱⲫⲏⲧⲏⲥ: ⲕⲉϣⲧⲁⲙⲉⲣⲱϣ: ⲉⲛ
ⲟⲩⲛ ⲛⲉϩⲟⲟⲩ ⲧⲏⲣⲟⲩ ⲡⲉϥⲱⲛⲁϩ

ⲟⲩⲭ ϩⲓ ⲁⲣⲭⲏ ⲑ ⲗⲓⲃⲁⲛⲟⲥ ϩⲁⲗⲟⲩⲑ ⲉⲣⲉⲡⲟϩ ⲙⲟⲩϩ ⲕⲁⲗⲟ

ⲥⲁⲗⲁⲑⲓⲏⲗ

ⲁⲡⲕⲟⲩⲧⲱⲛ: ⲥⲓⲃⲉ:
ⲡⲉⲭⲏϥ: ⲡⲧⲏⲥⲓⲱ:
ⲉⲛⲥⲉⲗⲁⲃⲁⲩ: ⲡⲓⲡⲣⲉ
ⲍⲁⲗⲏⲩⲅⲉ

ⲣⲉⲥ
ⲙⲟⲩϩ

ⲙⲓ

ⲡⲥⲁⲩⲱ ϯⲡⲁⲣⲁⲕⲁⲗⲓ ⲙⲙⲁⲕ ⲙⲡⲟⲟⲩ: ⲥⲁⲗⲁⲑⲓⲏⲗ
ϣⲱⲡⲉ ⲛ̄ⲁ ⲛⲉⲧⲉⲛⲁⲙⲓⲥ: ⲉⲧⲉⲛⲁ ⲓ ⲛⲉⲩⲣⲁⲛ
ϩ: ⲃⲁⲣⲁϩ: ⲥⲁϩ: ⲑⲁϩ: ⲡⲉⲛⲧⲁϥ: ⲃⲱⲕ: ϣⲁⲉⲩ
ⲥⲉⲛⲉϩⲟⲩⲛ ⲉⲡⲭⲁⲗⲭⲉⲗⲁⲃⲁⲡⲁⲧⲟⲛ ⲡⲉⲥ
ⲩⲥ: ϣⲁⲛⲧⲉⲥⲟⲩⲱⲙ ⲉⲃⲟⲗ: ⲡϣⲩⲛⲏⲁ: ⲛⲉⲥⲕⲱ

ABOVE Small papyrus codex (see page 28) containing simple magical symbols (2nd to 5th century CE). It is possible that small codices such as this were created as protective amulets rather than as information records. It has also been suggested by scholars that small papyrus and parchment codices with occult symbols or Christian prayers may have become amulets, transcending their original purpose.

RIGHT Papyrus codex (see also page 40) (6th century CE). Possibly the work of a Coptic priest, this codex contains Christian prayers, exorcisms, and charms to be copied out for protection. One of them calls upon God to protect the bearer from sorcery, incantations, and villainous people motivated by jealousy and envy. Wherever this written prayer was deposited "you [God] must guard the entrance and the exit, and all his dwelling places, and his windows, and his courtyards, and his bedrooms."

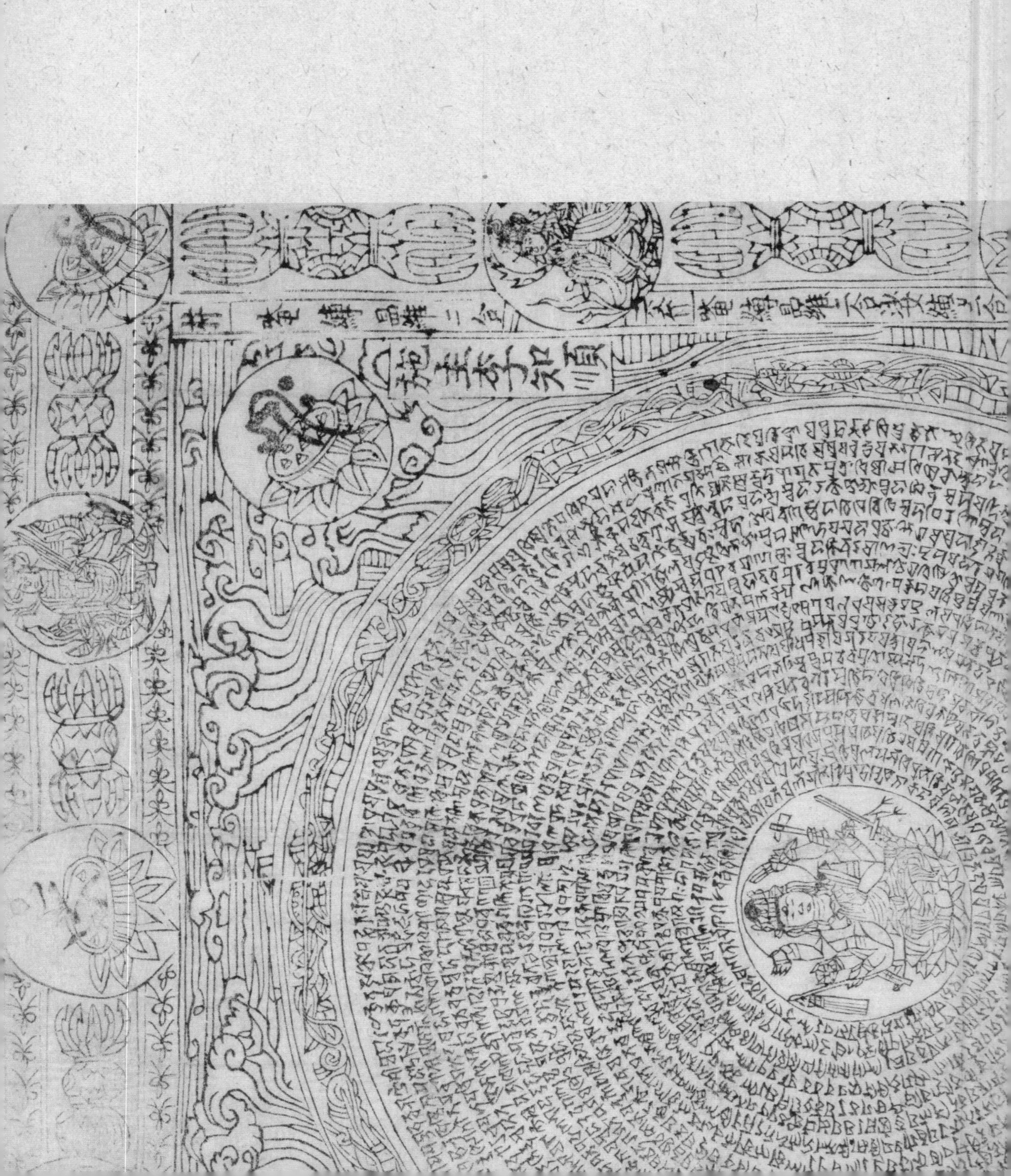

CHAPTER TWO

Parchment, Paper, and the Book

The magical texts produced in Coptic Egypt during late antiquity demonstrate the growing influence of parchment and paper in the expansion of written magic during the second half of the first millennium. Parchment, which is made from treated animal hides, began to be used increasingly from the 5th century in the West and the use of paper developed in the second half of the millennium in the Eastern and Arabic regions. Parchment and early paper were highly durable writing surfaces that enabled both the production of books as we know them today and ephemeral literary talismans written in ink that could be mass produced and easily manipulated.

The sacred Buddhist Mogao caves in China, which are located on the Silk Road near the western Chinese town of Dunhuang, are known for their exceptionally rich wall paintings and statues. Yet, one of the sealed-up grottoes (now known as the Library Cave) also contained an exceptional archive of thousands of early paper and silk documents dating from the 5th to the 11th centuries CE. Among the many *sutras* and prayers, and cultural and administrative documents, were a range of medical, magical, and divinatory texts, including numerous paper *dhāranīs*, or Buddhist amuletic mantras. They were written mostly in the Indian Brahmi script (the main vehicle for writing the Sanskrit language, which dates back to the 3rd century BCE), although Chinese would come to dominate as the main scribal language for *dhāranī* amulets by the end of the first millennium CE. The use of Brahmi in *dhāranīs* long outlived its use for secular purposes because it was considered a pure script in magical terms, and its foreignness also made it seem more potent.

With the fall of the Roman Empire (between 395 and 476 CE) there followed several centuries of distancing between western Europe and the eastern Mediterranean region, where Constantinople became the center of the Eastern Roman Empire. The great philosophical, medical, scientific, and magical writings of the Greeks, Egyptians, and Romans were largely lost to western European knowledge until the great flourishing of intellectual co-operation

between Jewish, Islamic, and Christian scholars in Spain and southern France during the 11th and 12th centuries. Much of the learning accumulated in the Graeco–Roman era was retained in Islamic and Jewish scholarship, where parchment was crucial to facilitating the transmission of this knowledge across long distances, with voluminous stitched-together manuscripts carried in saddlebags and trunks. Toledo was one of the key centers for the translation and exchange of this old knowledge, as well as the generation of new philosophical and scientific ideas across faiths and cultures. As a consequence, the city accrued a notorious reputation for being a center of magic.

Astral magic was the most influential Arabic occult conception to be widely adopted by Jewish and Christian magicians during the 11th and 12th centuries. This was the idea that everything on Earth was influenced by the stars, planets, and their associated spirits through invisible emanations from the heavens. The goal of the magician was to harness these heavenly powers by capturing them at astrologically propitious moments and locking them into talismans, or sigils, by the ritual use of word and image. In conception, the practice of astral magic was heavily influenced by beliefs and practices from antiquity, but it was novel and exciting in the medieval and Renaissance West, and handbooks emerged containing an array of sigil images and instructions on how to use them. Once the astral power had been captured, the talismans could be worn or buried for health, protection, provoking love, and also for enacting magical revenge. The most influential and sophisticated of these astral talismanic manuals was the *Ghāyat al-Ḥakīm* (*The Goal of the Sage*), first written down in the 10th century and then translated into Spanish, three centuries later, under the title *Picatrix*.

BELOW AND LEFT Pages from a Tibetan Book of Spells found in the Mogao Caves, Dunhuang, China (9th to 10th century). Written in ink on recycled paper, the title page states: "This is the ritual manual of Bhikṣu Prajñāprabhā." Within the text is a wide range of spells, including those for commanding demons, finding treasure, making friends, curing illnesses, pacifying evil people, and curing madness.

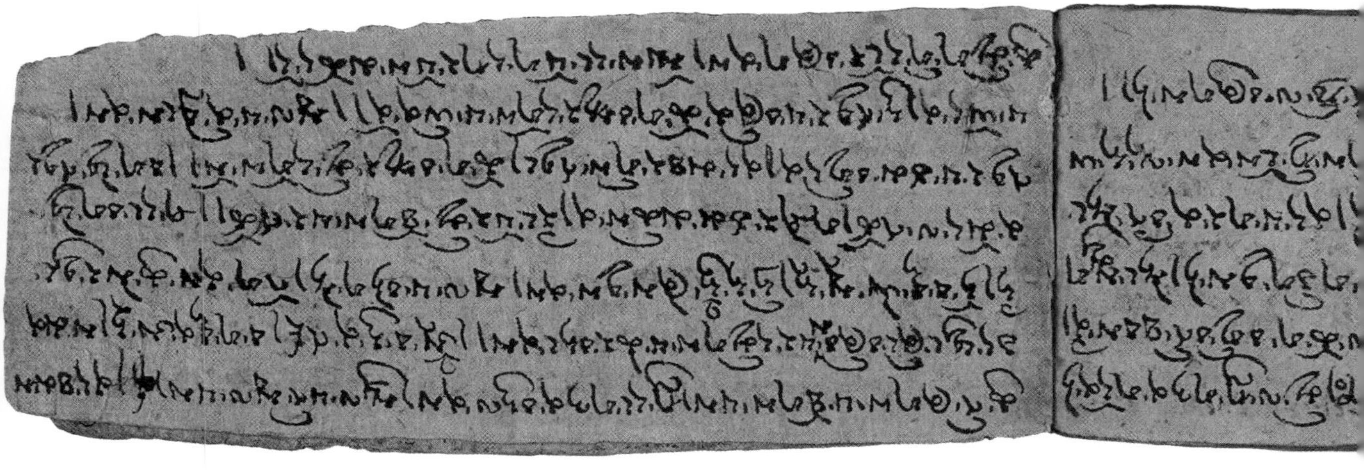

In pre-Islamic and Muslim Arabic beliefs there were various classes of demons but foremost among them were invisible beings, known as *jinn*, who had their own society and tribes. According to the Qur'an, *jinn* could be good as well as evil but they were certainly untrustworthy and usually unwelcome. They were also believed to be the cause of numerous illnesses and could be vengeful. Magic was one way of keeping the *jinn* away and negating their harmful influence on body and home. The Sufi cleric Abū al-Faḍl Muḥammad al Ṭabasī (d.1089), who lived in Nishapur, northeastern Iran, became notorious for his legendary command over the *jinn*. There is a story that the renowned Islamic scholar al-Ghazālī (d. 1111) asked Ṭabasī to call up the *jinn* before him, and they duly appeared like shadows on the wall. But when al-Ghazālī asked to speak to one of them, Ṭabasī replied, "You are not capable of seeing more of them than this." Ṭabasī's reputation stemmed from the circulation of one of his works entitled "The Comprehensive Compendium to the Entire Sea," which contained incantations, talismans, and spells for controlling demons and the *jinn* by legitimate Islamic rituals rooted in pious motives and pure means.

While the concept of *jinn* had little direct influence on western magic books in the medieval period, all three major religious magical traditions shared a preoccupation with angels and angelic communication. The word "angel" has ancient origins and referred to a "messenger" from the heavens. According to the hugely influential Christian theologian Thomas Aquinas (1225–1274), they were bodiless spirits created by God at the same time as he created the material world. They were intellectually superior to humans, knew all that passed

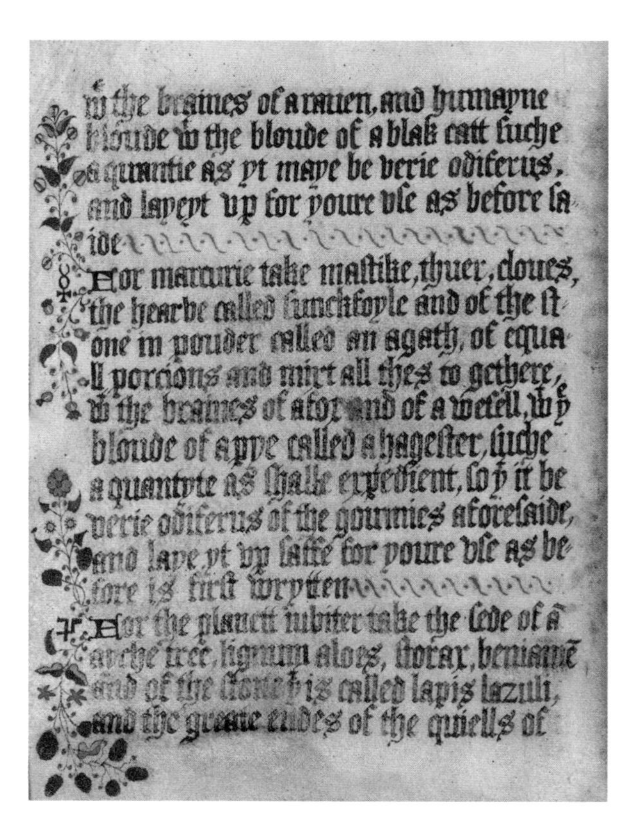
wt the braines of a rauen, and humayne
bloude wt the bloude of a blak catt suche
a quantie as yt maye be verie odiferus,
and kepeyt up for youre use as before sa
ide
For marcurie take mastike, thuer, cloues,
the hearbe called sinckfoyle and of the st
one in pouder called an agath, of equa
ll porcions and mixt all thes to gethere,
wt the braines of a fox and of a wesell, wt ye
bloude of a pye called a hagester, suche
a quantyte as shalle expedient, so yt it be
verie odiferus of the gummies aforesaide,
and laye yt up saffe for youre use as be
fore is first wrytten
For the planett iubiter take the sede of a
asshe tree, lignum aloes, storax, beniamine
and of the stone yt is called lapis lazuli,
and the greate endes of the quills of

RIGHT Coptic Christian amulet on vellum, Egypt (8th century). Magic symbols are depicted at the bottom, and above are the beginning words of each of the four Gospels, starting with the line: "The Holy Gospel according to Matthew. The book of the generation of Jesus the Christ, the son of David."

LEFT Page from a version of the *Sworn Book of Honorius,* or *Liber iuratus Honorii* (15th century), on parchment in English and Latin. This section describes a cleansing suffumigation in relation to Mars, which includes (at the top of the page) "the braines of a raven, and humayne bloude, with the bloude of a blak catt, suche a quantie as yt maye be verie odiferus."

in the world, and were in the service of God to aid humans to achieve virtue and grace. In Europe through the 13th and 14th centuries, several magic books of dubious authorship emerged, which provided monks and the learned elite with instructions for invoking angels in order to obtain this heavenly knowledge that was otherwise hidden from mere mortals. One influential text was the 13th-century *Sworn Book of Honorius*, purportedly written in the 4th century by one Honorius of Thebes. He was a legendary great magician, who had recorded the collective wisdom of the most powerful masters of magic from Thebes, Toledo, Naples, and Athens when they all met at a great council presided over by Honorius. The most widely copied work of angel magic, though, was the *Ars Notoria* attributed to Solomon. It contained a series of prayers and adjurations that often mixed Greek, Latin, and Arabic languages, and which were sometimes accompanied by illuminated illustrations of the angels. Although it was presented as a pious work, the *Ars Notoria* was condemned by some clergy.

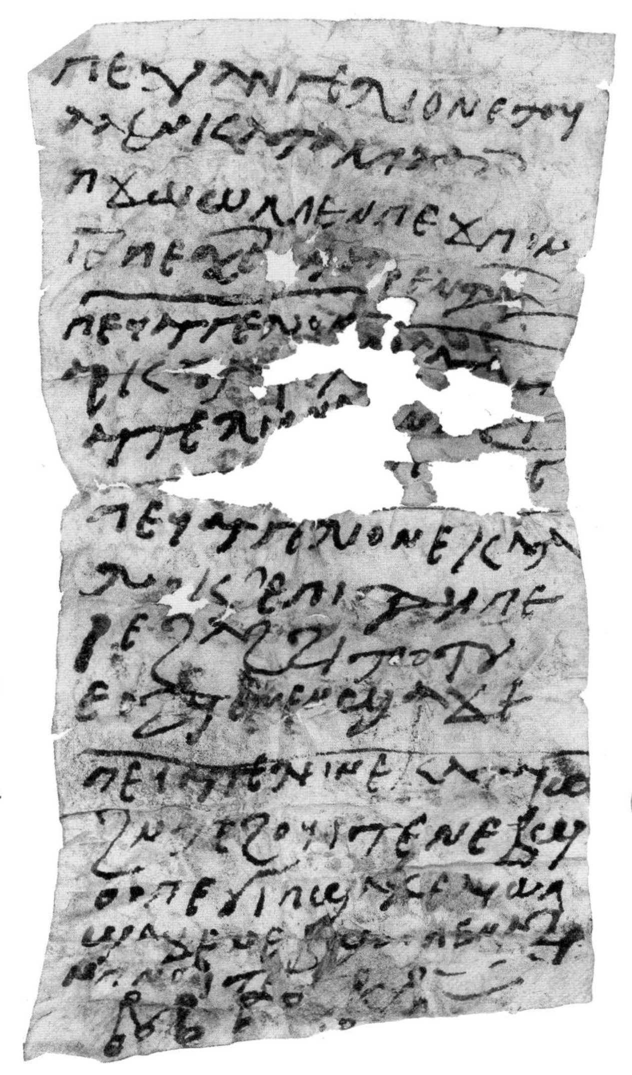

Just as there were good angels there were also fallen angels, whose aim was to corrupt humans and lead them down the path of damnation. In medieval Christian theology these demons were pure evil, unlike the more equivocal *jinn*. To use necromantic magic to contact or control them, usually for base human desires of wealth and power, was not only dangerous but also blasphemous. The increasing attacks against learned magic from the 13th century onward focused on the conviction that when magicians attempted to contact angels they were actually opening up portals for demons, who were masters of illusion and who could imitate divine spirits. For the likes of Thomas Aquinas this threat was sufficient to condemn all attempts at conjurations and adjurations and not just necromancy. As the 14th-century Benedictine monk, John of Morigny, said after long experimentation with the *Ars Notoria*, "It was twice revealed to me by all the angelic spirits that in this book's prayers in outlandish tongues there was an invocation of malign spirits hidden so subtly and ingeniously that nobody in the world, however subtle he be, would be able to perceive it."

NATURAL MAGIC, SCRIPTS, AND SYMBOLS

Not all medieval magic involved rituals and communications with the supernatural. Natural magic concerned the secret or occult properties of animals, plants, gems, and minerals.

It was seen as part of God's plan to imbue things with hidden potencies for the potential benefit of humankind. For some, He left signs or signatures of their properties, such as the magnetic qualities of lodestones or the resemblance of certain plants to parts of the human body that they could heal. But the secret properties of many things required scientific investigation, and the ancient Greeks and Romans had fortunately bequeathed a large body of findings for the benefit of medieval physicians and magicians. During the 12th century, the quest to know the secret power of language also developed with the rise of Jewish Kabbalah in Spain and Provence. Central to this mystical tradition was the notion that Hebrew had a divine origin and the Torah was, therefore, a repository of potent secret names for God that could enhance the power of prayers and talismans. The letters of the Hebrew alphabet were also imbued with mystical properties through their connection to God. It is not surprising, then, that Hebrew words and letters also came to be integrated into Christian books of ritual magic, and they were particularly prominent in the later works of Renaissance magic.

What gives medieval magical texts their distinctive artistic quality are the many figures, characters, signs, sigils, and seals they contain. Circularity was a dominant theme: sometimes circles represented cosmological ideas about the heavens, but they were also magical figures to

LEFT Watercolor and ink on paper, an illustration of a snake from a version of the ***Kitāb al-Manāfi' al-Ḥayawān*** (***Book on the Usefulness of Animals***), Iran (c. 1300). It was originally written by the Persian physician Ibn Bakhtīshū (d. 1058), who drew upon ancient Greek medicine for his manual on the medical properties of animals.

LEFT A diagram of spheres of the elements and planets rotated by four angels (early 14th century), on parchment in old French. This scheme illustrating how the medieval cosmos worked, shows Earth at the center, then the "wandering stars," including the Moon, Sun, Mercury, and Mars. Finally, there is Heaven and the "fixed stars" that appeared not to move in the night sky.

be drawn in the earth or on large pieces of parchment to protect the magician from spirits or to bind spirits within them. Such circles usually contained sacred words, the names of spirits and angels, and religious symbols such as the cross. From the 15th century, there was also an increasing representation in grimoires of triangles, pentagrams, and pentacles. The latter were particularly associated with Solomonic works, and in a Christian context represented either the seal of God, the five wounds of Christ, or the five senses. This sacred geometry would become integral to a variety of magical and esoteric traditions that emerged over ensuing centuries. The range of occult symbols were obviously essential to the practice of medieval magic but they also served an aesthetic function. The creators of such images often made use of color in artistic ways, which helped to attract the eye and intrigue the reader, thereby adding to the mystique of magical texts.

As well as the penchant for pseudepigrapha (falsely attributed works, such as the *Sworn Book of Honorius* and *Ars Notoria*, see page 57) in medieval magic, there was also a tradition of pseudo scripts. During late antiquity and the early medieval period, there are increasing examples of made-up languages in magical communications, consisting of words and symbols

ABOVE Illustration from an astral magic manuscript entitled *Astromagia* (c. 1280), on parchment. The scene shows a magician sacrificing a goat to the god Mercury (depicted riding on a peacock in Scorpio), and a brazier burning incense in the background. A compilation manuscript in the Vatican Library, *Astromagia* also contains a series of astral talismans, with clear influences from Arabic magic, including the *Picatrix* (see page 55).

that look like a recognizable language and yet have no discernible meaning. These can be found on curse tablets and binding spells, for example, and also on Arabic and Hebraic incantation bowls, along with *voces magicae,* or secret "magical names." The tradition of using pseudo-Hebraic language emerged again in medieval European magic books, where letters and words were mixed with divine names, symbols, circles, and pentacles. The 15th-century German grimoire known as the *Munich Handbook*, which was written in Latin and was used for necromantic purposes, borrowed quite heavily from earlier Jewish magic manuscripts. It contains a section referencing the *Semiforas* (*semhemphoras*, a mystical name for God), which included passages that combine Hebrew and pseudo-Hebraic words. In Nordic medieval studies we also find debate about the interpretation of pseudo-runes on runic pendants for health and protection. Arabic-looking pseudo scripts can also be found in medieval magic books. A Moorish Spanish handbook known as the *Libro de Dichos Maravillosos* includes instructions for a written charm to restore the love between husband and wife that required the drawing of some Arabic-style letters in raven's blood: "And you shall take its blood, and you shall write this writing. And you shall take its heart, and dry it, and crumble it, and you shall give it in a drink to the woman."

In the same period, pseudo script in the Daoist tradition of *Fu* talismans for healing, protection, and exorcism has been found on pottery, metal, wood, and paper. One origin story for the *Fu* script stated that it was created from the condensation of clouds. The talismans themselves were not necessarily intrinsically potent, as instructions survive noting how they needed to be activated by spells or ritual hand gestures. Their production and use flourished in the Song period (960–1279), and they were probably produced by a combination of astrologers, medical men, and monks. These talismans were variously carried on the person, placed in or under dwellings or tombs, and sometimes burned to ashes and ingested. The *Fu* tradition also found its way into Buddhism and was subsequently incorporated into Japanese *Shinto* and *Shugendō* religious cultures from around the 15th century.

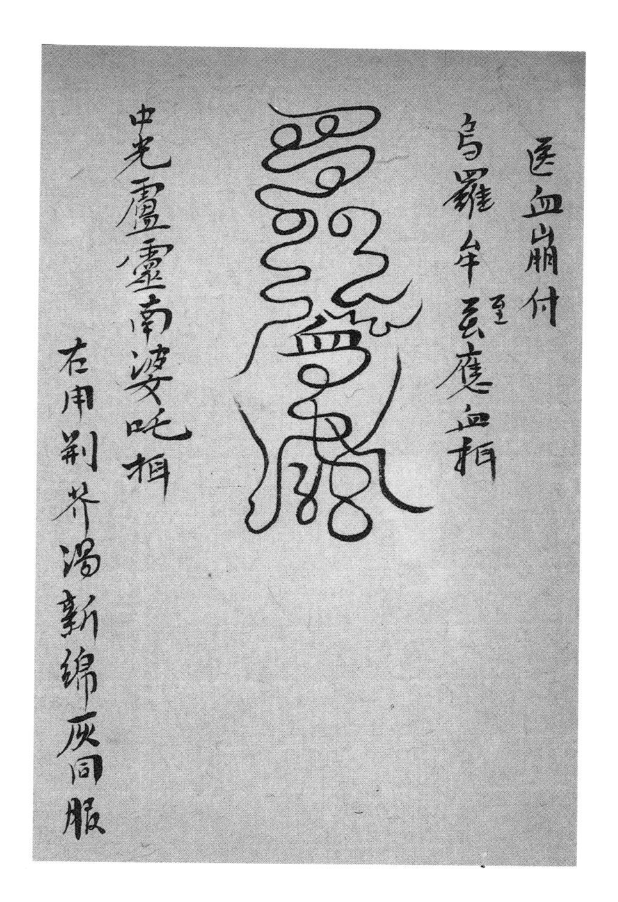

ABOVE Chinese *Fu* medical talisman, ink on paper. Used to cure menstrual flooding, this is from a manuscript copy of the *Tian Yi fu lu* (*Record of Spells of the Celestial Physician*). This anonymous book of spells and medical recipes appeared in the early 12th century, during the Northern Song dynasty.

It has been argued that pseudo scripts were the creation of illiterate charlatans, due to some of the unreadable texts on incantation bowls. Yet, this is unlikely—these scripts clearly had a meaning and purpose, even if it was only to mystify others. Besides, the act of writing was magical in and of itself and not just a means of transmitting magical knowledge. Some pseudo scripts were created as languages of the gods or spirits, so were not meant to be understood by mere mortals. Their secrets were only decodable by magicians, seers, and holy men. The "Celestial Speech," or Enochian language of the angels, recorded in private journals by the 16th-century English mage Dr. John Dee (1527–1608) falls into this category. Dee believed this angelic language, which was delivered through his scryer or medium Edward Kelley (1555–1597/8), derived from the time of Adam, who had used it to name all the things then known.

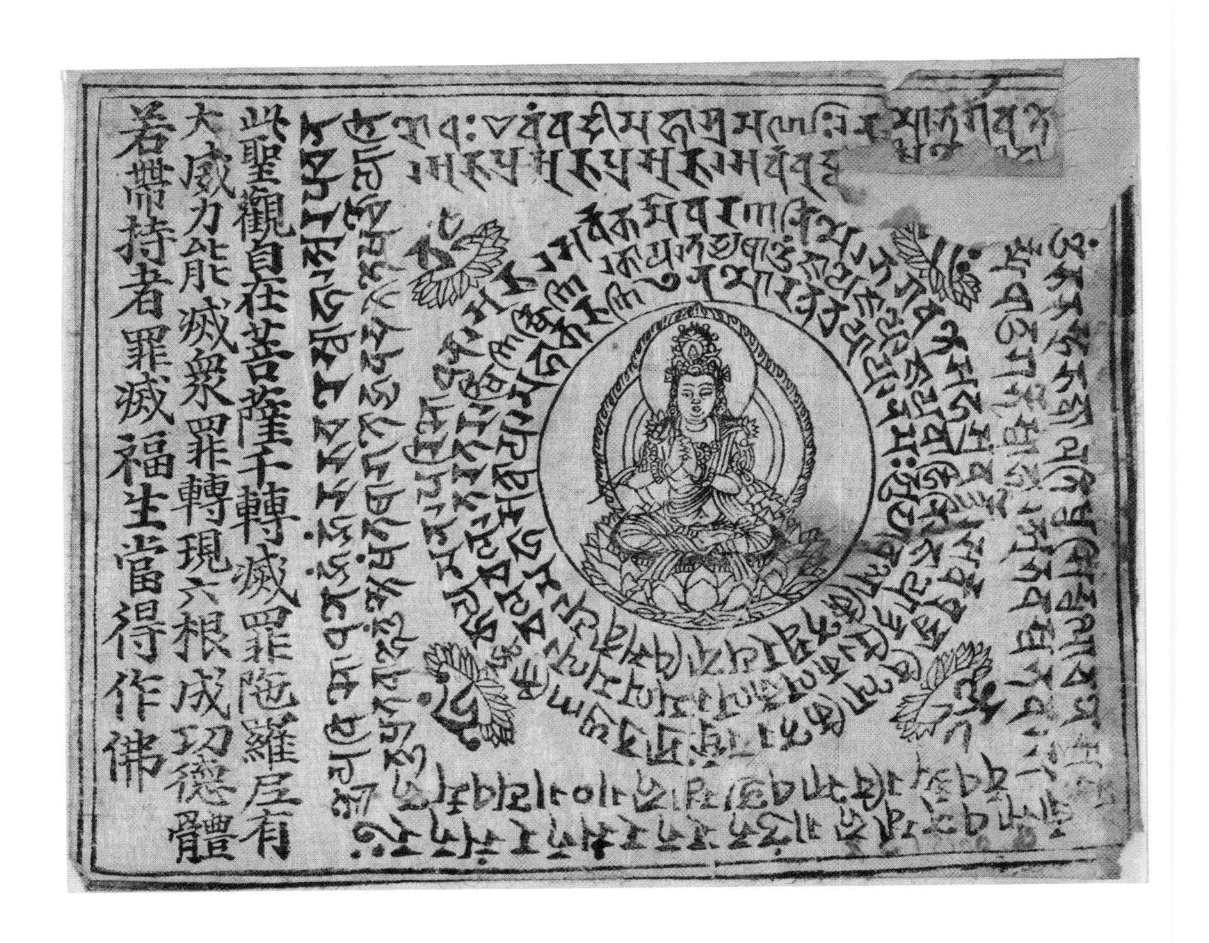

ABOVE Avalokiteśvara *dhāranī* (10th century). Found in the Dunhuang caves, China, this is a woodblock print of ink on paper. At the center is a line drawing of Avalokiteśvara, the hugely popular figure of Buddhist legend who is the earthly manifestation of the eternal Buddha. More to the point, he guards the world, and his compassion knows no bounds. Hence, he was (and still is) a popular figure to include in protective talismans. Around his figure is the Sanskrit *dhāranī*, or mantra, to be recited repeatedly. A Chinese translation of the mantra is also provided at the side so that lay people could use it for protective purposes.

RIGHT *Dhāranī* woodblock print of ink on paper (980 CE, Northern Song dynasty), produced in Dunhuang. At the center of this large protective sheet is an image of the eight-armed goddess Mahāpratisarā meditating in a yogic pose. She is a great protectress represented in statues, stelae, and art. Around the image is the *dhāranī* written in Sanskrit, with a Chinese translation provided at the bottom. Images of lotus flowers, which represent the purity of body and mind, and spiritual awakening, are positioned around the text.

施主李知順
王文沼雕板
大隨求陀羅尼
若有受持此神呪者所
在得勝若有能書寫帯
在頭者若在臂者是人
能成一切善事最勝清
淨常為諸大龍王之所
擁護又為諸佛菩薩之
所憶念此神呪能与衆
生最勝安樂不為夜叉
羅刹諸鬼神等為諸惱
害亦不為寒熱等病之
所侵損厭蠱呪詛不能
為害先業之罪受持消
滅持此呪者常得安樂
無諸疾病色相熾盛圓
滿吉祥福德增長一切
呪法皆悉成就
若有人受持供養切
宜護淨 太平興國五
年六月二十五日雕
板畢手記

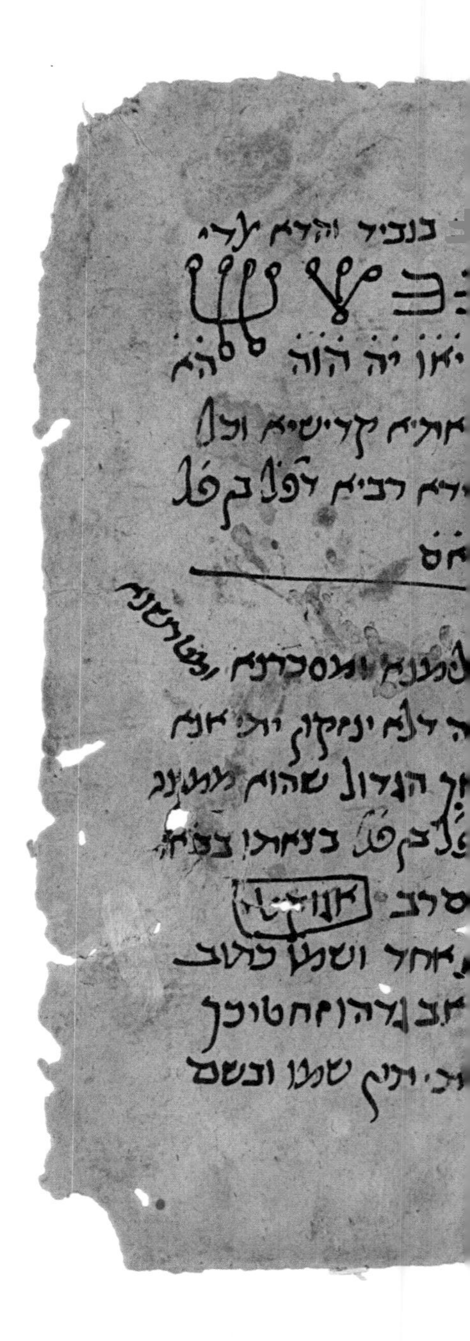

LEFT Paper amulet written in Hebrew, Judaeo-Arabic, and Aramaic (date uncertain, but probably 11th to 13th century). This was produced to protect a woman called Esther from demons, sickness, and the evil eye. This amulet was found in the Cairo *Genizah* (Hebrew for "hiding place"), the storeroom of disused religious texts of the Ben Ezra Synagogue, which contained tens of thousands of manuscript fragments dating from the 6th to 19th centuries.

ABOVE Recipes from a book of magic written in Hebrew, Aramaic, and Judeo-Arabic (11th to 13th century). They include spells to make peace between a man and his wife, to silence an enemy, and to abort "a dead fetus in his mother's womb." At the top of the left-hand page is a spell to release someone from a binding spell, in other words to remove an impotence curse. At the bottom of the page is a charm "For blocking one's mouth," in order to prevent an act of witchcraft from taking effect.

Making Parchment and Paper

Made from animal skin—usually that of sheep, goat, or calf, but also horse, camel, and deer—parchment was created by scraping, bleaching, and drying the skins under tension on a frame. Although it is generally thought to have been invented in the 2nd century BCE, simple animal skins were used for writing much earlier. The finest quality parchment was known as vellum, and was made from the skin of unborn or young calves. Grimoires sometimes instructed that such virgin or unborn parchment be used in the creation of amulets because of its intrinsic purity, while Jewish tradition forbade the use of pig or camel skins for talismans. Parchment was more expensive to produce than papyrus, and it has been estimated that the 800-page *Codex Sinaiticus*, a 4th-century Christian Greek Bible, would have required hundreds of animal skins.

Yet, parchment created the book as we know it today: it could be easily cut, stitched, and bound together, and it was dense enough to allow writing on both sides of a page without ink bleeding through. This made it a lighter, more compact, and durable record of knowledge—particularly in the damper climates of Europe—than papyrus scrolls. Parchment was a revolution, not just in writing technology but also in transferability and portability—the written word was now on the move as never before. In the late Roman era, a new type of stable ink was also invented, based on ground iron sulfates mixed with gum Arabic and the tannin from gallnuts. This acidic mixture became the standard black ink for writing with a quill on parchment in medieval manuscripts, where the red iron tinge can often be seen today.

LEFT Medieval fresco from the Church of Saint George at Reichenau Abbey, Lake Constance, southern Germany (14th century). It shows four demons stretching out a whole, uncut parchment hide. Another demon is depicted writing on the parchment regarding the fate of the foolish gossiping figures just above.

ABOVE Drawing of a parchment maker at work, from *Hausbuch der Mendelschen* (*Mendel's House Book*), on paper (c. 1425). The parchment maker stands in front of a large stretching frame, and works the animal skin with a specialized crescent-shaped scraper to remove hairs and imperfections. It was important to dry the skin under tension.

Although the word "paper" derives from "papyrus," paper was actually invented by the Chinese in around the 1st century CE. We know of a high-ranking court official named Cai Lun who, in 105 CE, started the mass manufacturing of paper for official documents, using mulberry bark mixed with hemp waste, old fishing nets, and rags. These were soaked until the individual fibers could be teased apart. Then, the wet fibers were placed on a screen of cotton or hemp, and washed and strained so that it created an even, smooth surface ready for drying. It was likely through trade along the Silk Road, as well as military encounters, that paper-making spread to the Arab world. Paper mills were set up in Baghdad in the 8th century and began to spread westward from there. The first evidence for paper-making in Europe was in 11th-century Islamic Spain, with Toledo an early center of production, but it was not until the 14th century that it was adopted in western Europe. When made solely from old textile rags, rather than a specific type of plant, paper could be made anywhere, regardless of the climate.

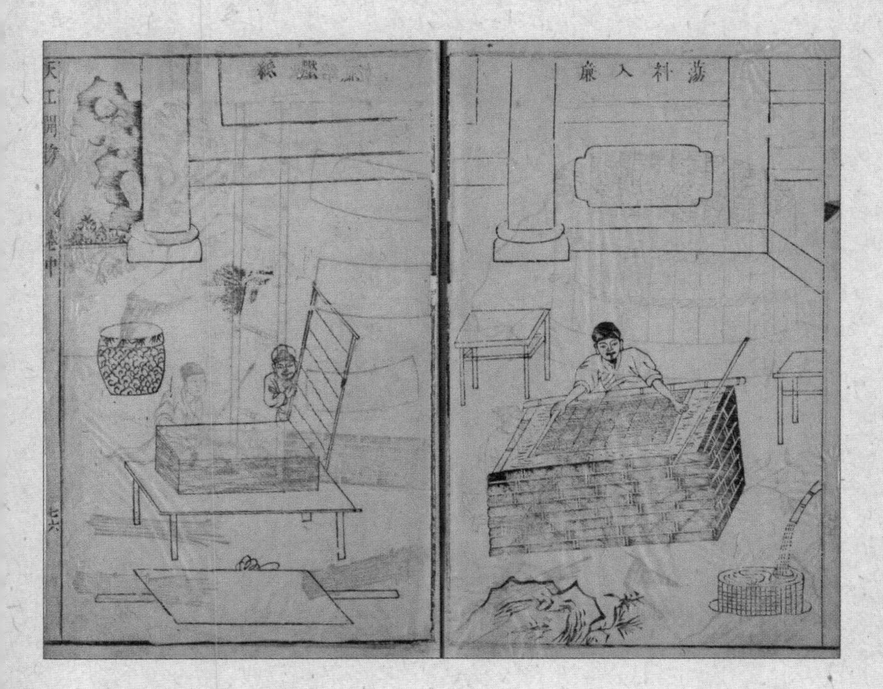

LEFT Illustrations from *Tiangong Kaiwu* (*The Exploitation of the Works of Nature*), depicting the final stages in Chinese papermaking, using bark, rag fibers, and bamboo (1637). This was compiled by Song Yingxing (1587–1666), a scientist and encyclopedist, whose book provides valuable insights into the secrets of manufacturing processes in China.

ABOVE Fragment of a magical text in ink on parchment (8th to 9th century), Egyptian. Little is known about the context of this fragment, but the figure resembles the sorts of winged demons and angel figures found on Coptic magical papyri. It demonstrates the transition of this written magic tradition to parchment in the period.

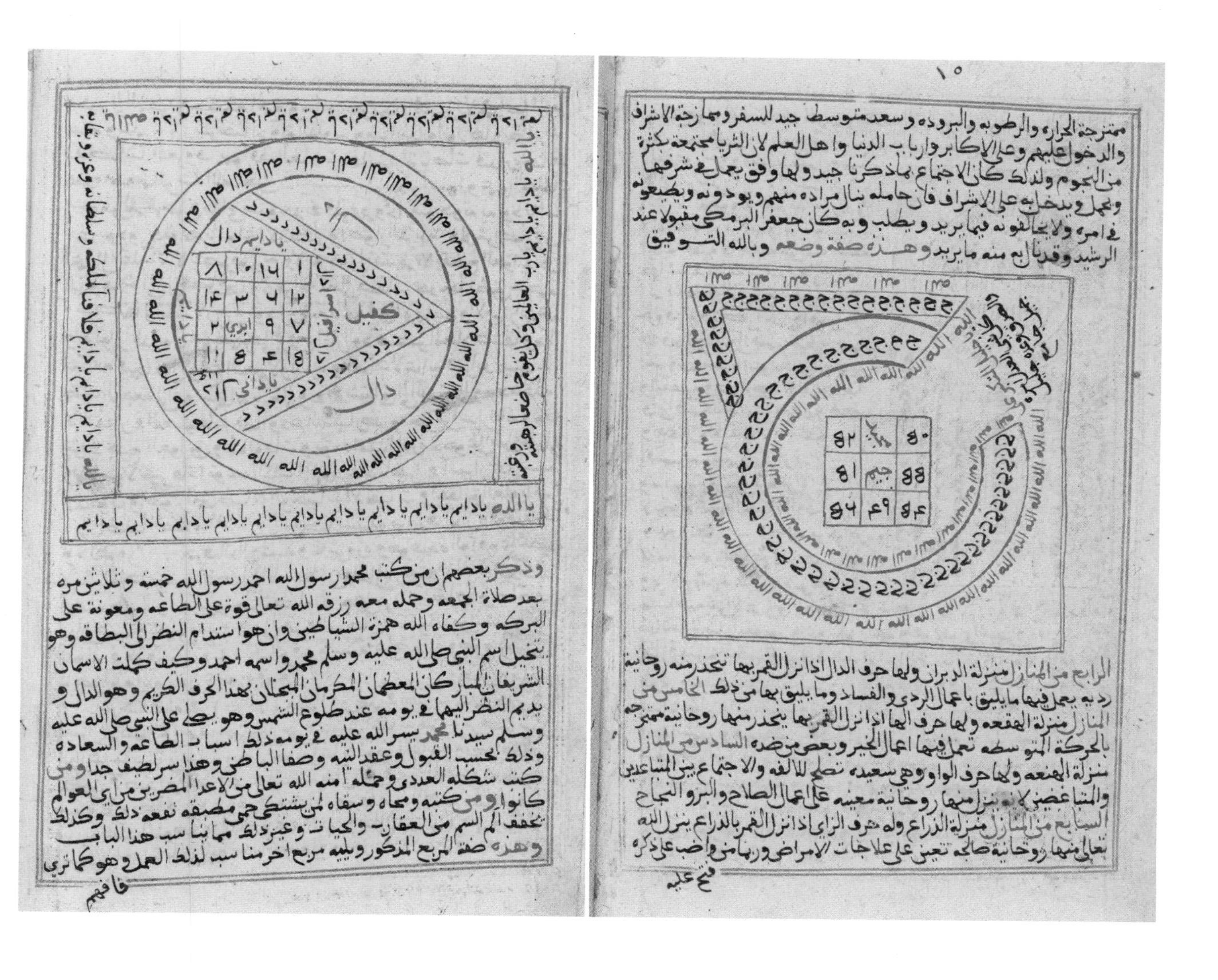

ABOVE Pages from a manuscript copy of Aḥmad ibn al-Būnī's *Shams al-ma'ārif* (*The Sun of Knowledge*)—a practical manual of Islamic and Islamicized magic written in Arabic on paper (early 13th century). It contained relatively little theory, and was full of recipes and instructions for creating talismans. Born in Algeria but based in Egypt, al-Būnī (d. 1225) was a Sufi scholar of mathematics, philosophy, and the occult. We know little of his life, yet, once copies of the *Shams al-ma'ārif*, and other of al-Būnī's related works (although some were spuriously ascribed to him), began to circulate, they had a significant influence on Islamic magic tradition and practice. As depicted in the designs above, al-Būnī argued that the only way to talk to or harness *jinn*, angels, and other spirits was through the potent combination of Arabic letters known as *ilm al-huruf* (the science of the letters), magic number and letter squares, the 99 "beautiful names of God," and occult geometry.

ABOVE Image from a manuscript copy of Abû Ma'shar al-Balkhî, *Kitâb al-mawâlîd* (*The Book of Nativities*), Egypt (14th century). Abû Ma'shar al-Balkhî (787–886), known in Western Europe as Albumasar, was a great Persian astrologer based in the Abbasid caliphate court in Baghdad. His numerous manuals on astrology and astronomy were translated into Latin, and had a deep influence upon intellectual astrology across the medieval Islamic and Christian worlds. The inspirations for his work were wide ranging, from ancient Greek philosophy to Indian sources. Here, the lion (Leo) is represented in association with the Sun, which is significant as there was a medieval Christian debate that the Sun lay in the constellation of Aries.

RIGHT Page from a version of Abd al-Rahmân al-Sûfi's, *Kitāb ṣuwar al-kawākib al-thābitah* (*Book of the Images of the Fixed Stars*), written in Arabic (13th century). Al-Sûfi (903–986) was a Persian astronomer based at the court in Isfahan, Iran, who translated and enhanced the astronomical texts of the Graeco-Roman scientists. *The Book of the Images of the Fixed Stars* was his amended version of the *Almagest* by the Greek astronomer Ptolemy (c. 100–170 CE), a complex work of mathematics on the motion of the stars and planets. Here, the image shows *Ursa major*, the Great Bear. The constellation is drawn twice in mirror image to depict how it was observed from the sky (using a celestial globe), and the earth as seen in the night sky.

وبين الثاني من الاثنين الخارجين عن الصورة قدام كبد الاسد وبين الذي على المابض كوكب من القدر الخامس من اصغره
هو الى الثاني الخارج عن الصورة اقرب ٥ وداخل الحوض كوكب هو مع السابع والثامن على مثلث ٥ وكوكب
بين التاسع والعاشر قد صار معهما على مثلث منفرج الزاويه ٥ وعلى جنوب القابد كوكبان من القدر
السادس بينهما في رأى العين ارجح من ذراع ٥ وبين القابد وبين الاقرب اليه من الاثنين نحو ذراع وهما
متقدمان له وهو يتلوهما لم يذكر شيئا منها وكذلك في خلال الصورة وحواليها كواكب كثيرة فيها
من القدر الخامس والسادس ٥ فاما الخفية الخارجة عن الاقدار الستة فهي بلا نهاية وجميع ذلك من جملة الطباء
الطباء واولادها ٥

صورة الدب الاكبر على ما يرى في الكرة ٥

صورة الدب الاكبر على ما يرى في السماء

جدول

King Solomon, the Magician

According to the Old Testament, Solomon was the son of King David and successor to the throne of the kingdom of Israel. Calculations from Biblical genealogies place his reign between 970 and 931 BCE. In the Bible, Solomon is celebrated for his great wisdom and for the construction of the great temple in Jerusalem that housed the Ark of the Covenant. There is no mention of Solomon the magic worker, though. So, how did his reputation as a magician come to be so pervasive as to partially outweigh his Biblical significance?

As well as the *Ars Notoria* (see page 57), Solomon was also attributed the authorship of a Greek text circulating in the early centuries of the Common Era known as the *Testament of Solomon*, which tells of his various dealings with demons and how he overcame them. It includes an account of Solomon's seal, or ring: Solomon prayed day and night in the great temple on behalf of a young man who was being plagued by a vampiric demon named Ornias. The archangel Michael appeared from the heavens and gave to Solomon a ring bearing the mark of God in the form of a pentagram (though later a hexagram was also used to represent the seal), that gave its possessor the power to command all demons. Over the centuries, copies of the *Testament* proliferated in Hebrew, Latin, and Arabic, cementing Solomon's role as a great magician across all three major religions in Europe and the Middle East.

Then, in the 15th century, a new Solomonic grimoire appeared, the *Clavicule*, or *Key of Solomon*, which would place Solomon further at the center of the European magical tradition. It came with

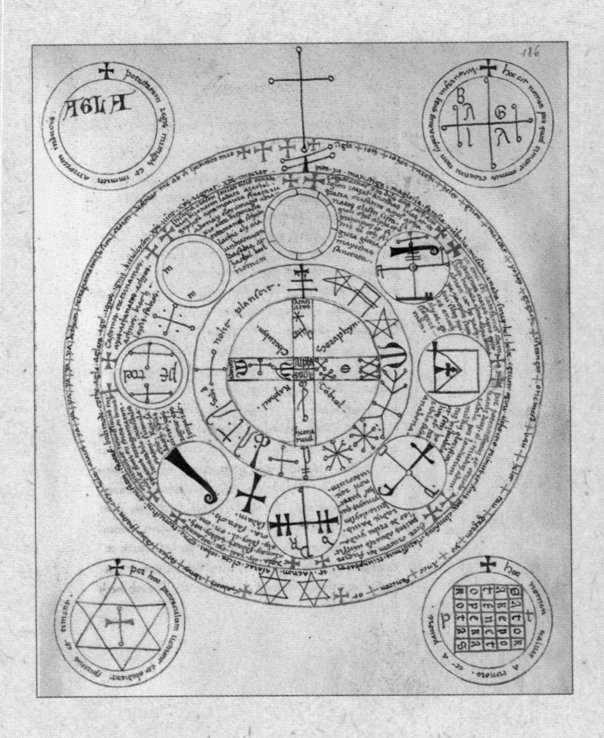

RIGHT Illuminated letter "P" depicting King Solomon, from a medieval Bible, Germany (late 12th century). We see Solomon writing his Proverbs, which are included in one of the books in the Old Testament. The Christian association of Solomon with great wisdom, as well as the building of the Temple in Jerusalem, was an enduring influence on Western esotericism, including Freemasonry.

OPPOSITE LEFT A highly elaborate Solomonic circle surrounded by four seals, from an Italian manuscript of the *Key of Solomon*, on parchment (1446). This is a very early vernacular copy of the *Solomonic* grimoire, linked to the Court of Milan. It demonstrates the graphic importance of circular magical imagery to the development of literary magic.

OPPOSITE RIGHT Depiction of King Solomon, or Sulaiman, in a miniature painting from a copy of the *Shāh-nāmeh* (*Book of Kings*) by the Persian poet Ferdowsi (c. 940–c. 1019/1025 CE), Iran (16th century). Solomon was recognized as a prophet in the Qur'an, and so plays a significant role in Islamic magic, particularly in relation to his seal ring and the control of *jinn* (see page 56).

its own discovery narrative, or "find story"—a fictional device common to grimoires and other esoteric texts from the ancient world to the present day. Solomon tells his readers that he wrote the *Key* for his son Rehoboam, and told him to conceal it in his tomb after his death. Many years later, some Babylonian philosophers embarked on repairing the tomb and discovered it. Versions of the *Key* that proliferated in the 16th and 17th centuries were full of practical magic, including illustrations and instructions for dealing with thieves, affairs of the heart, and how to become invisible, as well as the familiar "experiments" for conjuring spirits. There was no single version of the text, and it became a mutable vessel for magicians to copy down the practical magical knowledge of their desire, with the name of Solomon providing a venerable seal of approval.

ABOVE Page from the *Kitāb al-Bulhān* (*Book of Surprises*), Baghdad (14th to 15th century). Compiled by Abd al-Hasan Al-Isfahani, this Arabic manuscript is a compilation work that includes extracts from the likes of *The Book of Nativities*, as well as some fine astrological and *jinn* illustrations, perhaps by Al-Isfahani. Several of the vignettes in this image represent hybrid beings, which was a common concept in medieval Islamic cosmology. The angels of the seven heavens took on the guise of animals, such as the rooster-angel singing the praises of Allah, and strange "monstrous" human forms.

RIGHT Image from a copy of Abû Ma'shar al-Balkhî's, *Kitâb al-mawâlîd* (*The Book of Nativities*), Egypt (14th-century manuscript). The image depicts Târish, a king of domestic *jinn*. He is not listed in some other well-known books of *jinn*, but he bears a close resemblance to the more widely known "Red King" Al-Malik al-Ahmar, who was also depicted riding a lion and is associated with the planet Mars. It's suggested that this image represents Târish as a useful variant demon of the Red King, to be called upon to protect the home from snakes, hence biting into one snake and taming another to use as a leash.

شكل على الملك طارش ملك العمار واعو

ABOVE Illustrations from the *Kitāb al-Bulhān* (*Book of Surprises*), Arabic manuscript (14th to 15th century). The book presents the seven "Kings of the *Jinn*," each of them connected with a day of the week, a metal, an angel, and a planet. On the left, is the "Black King" Al-Malik al-Aswad, the *jinn* of Wednesday, with his marvellous retinue. On the right, is Zawba'a, the demon king of Friday, with his monstrous four heads, and his servant *jinn* around him. In legend, he is sometimes the embodiment of whirlwinds.

RIGHT Latin text entitled "Opera medica, astronomica et astrologica" (c. 1458–1459). This is an early Latin version of the *Picatrix* (see page 55), or *Ghāyat al-Ḥakīm* (*The Goal of the Sage*) from the Jagiellonian Library, Kraków. The page shows a couple of astrological charts of the 12 houses (each triangle represents an astrological house). The name of the person or event for which the horoscope has been drawn up is written in the square at the center, along with the date and hour. Such early Latin versions of the *Picatrix* were not direct translations from the Arabic but from an original Spanish translation (mid-13th century). By the mid-16th century, Latin editions of the *Picatrix* circulated around Europe, attracting numerous denunciations for being diabolic and "superstitious." This is likely why no printed editions appeared in the early modern period at a time when printed astrological treatises were abundant.

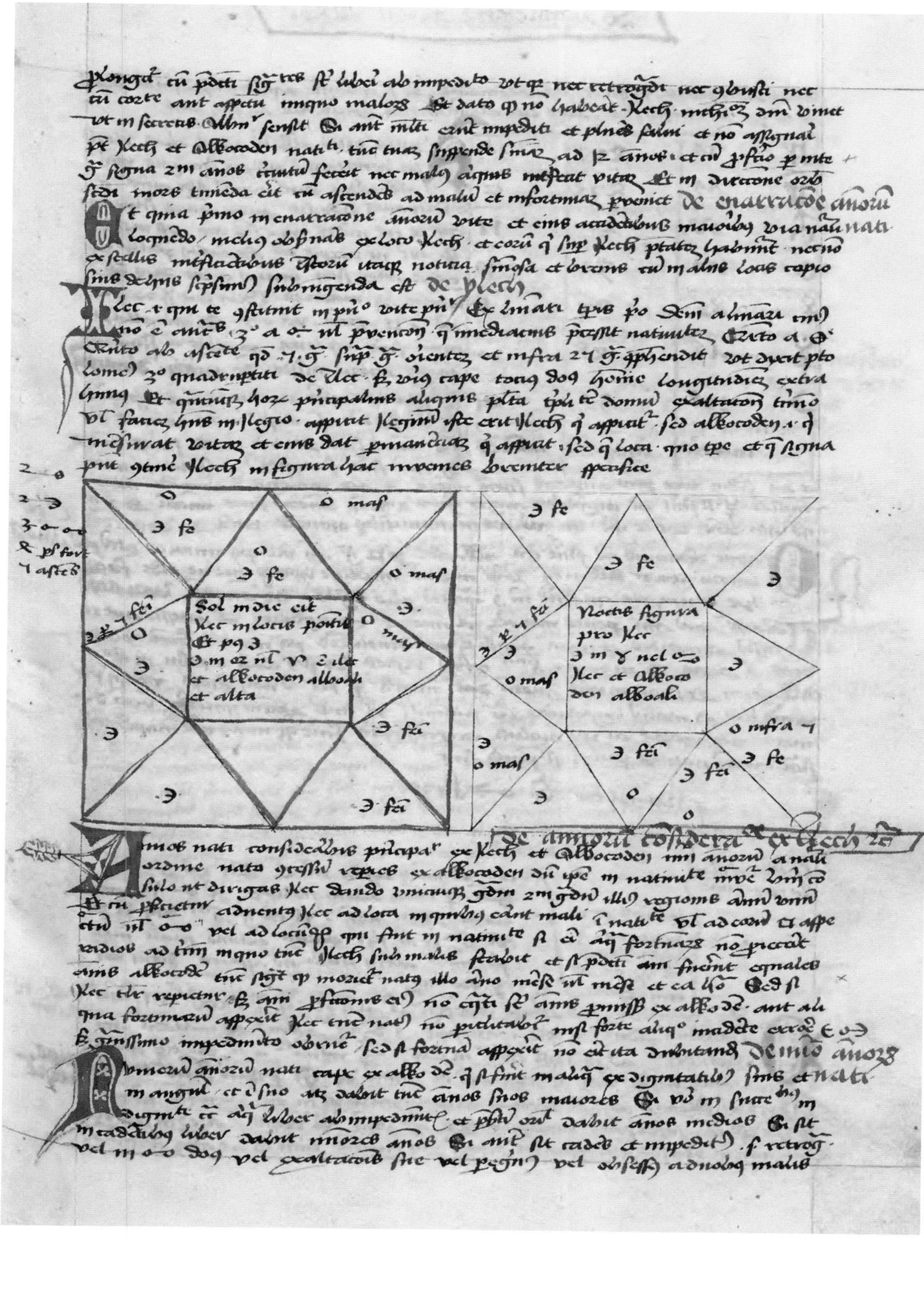

prolongat cum predicti sic [illegible] sit liberi ab impedimento vel quod nec retrogradi nec [illegible] nec
cum corde aut aspectu iniquo malorum Et dato quod non habeat hylech [illegible] diminuet
vel in secretis albu[mas]ar scripsit Et aut [illegible] erunt impediti et plures [illegible] et non assignar
pr hylech et alkocoden nativitatis tunc tua [illegible] sufficiente [illegible] ad [illegible] annos et cum [illegible] per [illegible]
signa [illegible] annos [illegible] fecerit nec malum aliquis infecerat vitam Et in directione [illegible]
[illegible] mors [illegible] cum ascendens ad malum et infortunam provenit De enarratione annorum
Et quia primo in enarratione annorum vite et eius accidentibus maioribus [illegible] nati
loquendo [illegible] ex loco hylech et eorum que super hylech potestatem habuerint necnon
ex stellis [illegible] itaque notitia [illegible] et [illegible] cum in aliis locis capio
[illegible] subiungenda est De hylech

Hylech [illegible] in principio vite principium Ex luminaribus [illegible] po [illegible] almazi [illegible]
[illegible] provenit [illegible] mediationis perfecit nativitatem [illegible]
[illegible] ab ascendente [illegible] super [illegible] orientem et infra 27 gradus [illegible] vel dicit ptolomeus
[illegible] quadripartiti de hylech [illegible] capere [illegible] domus longitudinem extra
[illegible] Et [illegible] principalibus aliquis [illegible] domum exaltationis [illegible]
vel faciem [illegible] in [illegible] appetit [illegible] hylech [illegible] sed alkocoden [illegible]
[illegible] vitam et eius dat permanentiam [illegible] sed que loca [illegible] et que signa
[illegible] hylech in figura [illegible] invenies [illegible]

2 o
2 [illegible]
3 o o
& ps fort
7 asted

Sol in die cum
Hylech in locis posuit
Et [illegible]
[illegible]
et alkocoden alboali
et alia

o mas
[illegible] fe
o mas
[illegible] fei

Noctis signa
pro Hylech
[illegible] vel [illegible]
Hylech et Alkoco
den alboali

o infra 7
[illegible] fe
[illegible] fei
o mas

De annorum considera[tione] ex hylech [illegible]

Annos nati considerabis principaliter ex hylech et alkocoden [illegible] annorum a nati
ordine nati gressum [illegible] ex alkocoden [illegible] in nativitate [illegible]
[illegible] ut dirigas hylech dando unicuique gradum [illegible] illius regionis annum unum
Et cum [illegible] advenerit hylech ad loca in quibus erant mali in nativitate vel ad eorum [illegible]
[illegible] vel ad loca [illegible] qui fuit in nativitate [illegible] fortunas non [illegible]
[illegible] ad [illegible] iniquo tunc hylech [illegible] malus [illegible] et si predicti [illegible] fuerint equales
cum alkocoden tunc [illegible] quod morietur natus illo anno mense vel mense et [illegible] Sed si
hylech [illegible] reperitur [illegible] non [illegible] ex alkocoden aut ali
[illegible] fortunarum aspexit hylech tunc natus non periclitabitur nisi forte aliquo incidente errore [illegible]
[illegible] gravissimo impedimento [illegible] sed si fortuna aspexit non erit ita dubitandum De numero annorum

Numerum annorum nati [illegible] ex alkocoden [illegible] fuerit in aliquo ex dignitatibus suis et [illegible]
in angulo et [illegible] dabit tunc annos suos maiores Si vero in [illegible] in
dignitate et [illegible] liber ab impedimento et [illegible] dabit annos medios Et si
in cadentibus liber dabit minores annos Si autem sit cadens et impeditus [illegible] retrog[radus]
vel in [illegible] vel exaltationis [illegible] vel [illegible] vel [illegible] malis

ABOVE Page from an abbreviated Hebrew version of the *Ghāyat al-Ḥakīm*, or *The Goal of the Sage* (late 15th to early 16th century). It is one of three, or so, surviving fragments from early Hebrew versions that had been translated directly from the Arabic by scribes in Italy. The Hebrew *Goal of the Sage* would have been particularly valued by students of the flourishing Jewish and Christian interest in Kabbalah during the Italian Renaissance.

RIGHT Angel blowing a celestial trumpet, an ink and watercolor painting from Iran (c. 1500). Angels played an important role in Islamic magic, along with *jinn* and *ruhaniyyat* (spiritual forces related to the planets). The names of the archangels were often inscribed around magic number and letter squares, and they acted as the guardians of certain days of the week. They could also be invoked through ritual. One such example is as follows: "Oh Bayel, angel assigned to the shining Sun, the benefactor of the world, of perfect light and luminosity, bringer of happiness and bad fortune, beneficial and harmful, by the name of the Master of the higher firmament, may you do so and so for me." The invocation was to be accompanied by the slaughter of a calf and the eating of its liver.

ABOVE "A guardian angel guides a man away from the Devil", from the *Livre des anges*, a French translation of the *Libre dels àngels* (1392) by Francesc Eiximenis (c.1340–1409), a Catalan Franciscan friar. The summoning and invocation of angels was an important aspect of Christian medieval magic, influenced heavily by Jewish and Islamic traditions. Knowing the hierarchy of angels, and their individual qualities, was essential to the practice of theurgy ("God work"), a means to obtain higher knowledge from the angels through liturgical ritual. According to some texts, there were certain celestial angels who could not be invoked at all because they served only God. Then there were the more amenable "terrestrial" and "aery" angels.

RIGHT Page from the *Sworne Book of Honorius* (*Liber iuratus Honorii*), in English and Latin (15th century). At the top, the angels of Mars are depicted, namely Samahel, Satyhel, Ylurahyhel, and Amabyhel. These angels, colored red "like burning coal," were responsible for provoking wars, and the death and destruction of men and animals. The golden angels of the Sun at the bottom are named as Raphael, Cashael, Daryhel, and Haurathaphel. They were the good ones, giving "love and favor and riches to a man and power also to keep him hail and to give dews, herbs, flowers and fruits in a moment."

Samahel · Satyhel · Ylurahyhel · Amabyhel ·

the sealle of the angells of mars Is thys

[sigils] : and there nature Is to cause & styre vp warre murder distructyon and mortalyte of people and of all earthly thinges, & there bodyes are of a meane statuer drye & leane there colouer Is redd lyke to burning colles burning redd, and there regyon or abydinge Is the sowthe

off the sprites that be under the planet of the sonne other there are under the sonne which are these

Raphael · Cashael · Dardyhel · Hanrathaphel ·

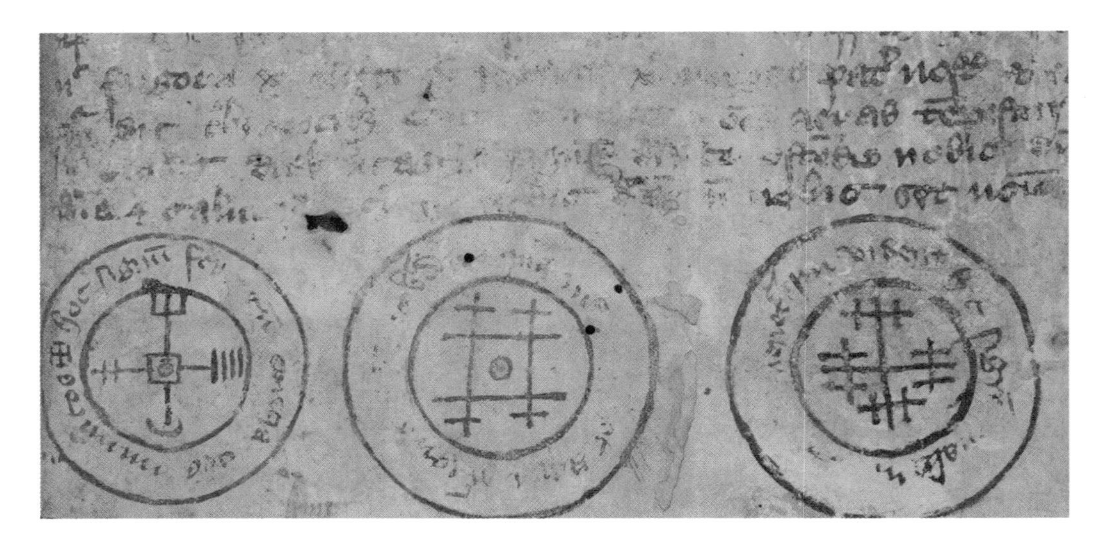

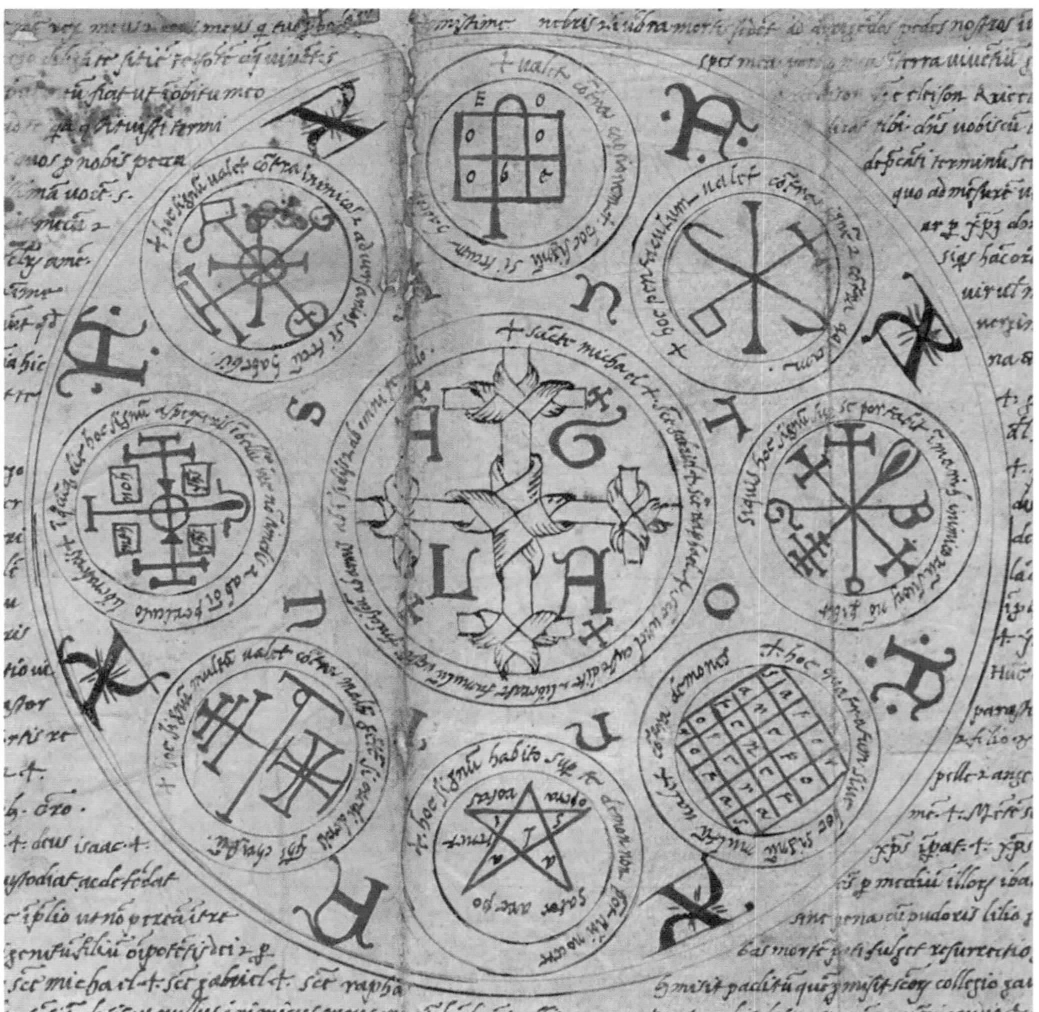

TOP LEFT Three magical seals added to a parchment *Book of Hours* (a Christian devotional text), England (15th century). The inscription on the first seal reads: "*Hoc signum fer [te]cum contra omnes inimicos*" ("Carry this seal with you against all enemies"). The other two seals were designed against flooding, fire, and evil. In the same hand on another page (not pictured here) is a "Seal of Solomon," with a hexagram to protect the wearer from being captured in battle.

BOTTOM LEFT An early 16th-century parchment amulet made for a man named Francesco, Italy. It consists of nine magic seals for different purposes, including protection against demons and misfortune. In between each seal are the Greek letters Chi (X-like) and Rho (P-like), which represent Christ.

ABOVE Two pages from a parchment manuscript entitled *Ars notoria, sive Flores aurei* (*Notary Art, or Golden Flowers*), France or North Italy (c. 1225). This is a variant of the Solomonic *Ars notoria* (see page 57) that was sometimes spuriously attributed to "his friend and successor," the Greek philosopher Apollonius of Tyana (c. 3 BCE –97 CE). It consists of a series of complex esoteric diagrams, which the magician was to meditate upon while reciting the accompanying invocations or prayers in sequence. It mixed the "science" of memory with liturgical devotion for magical purposes.

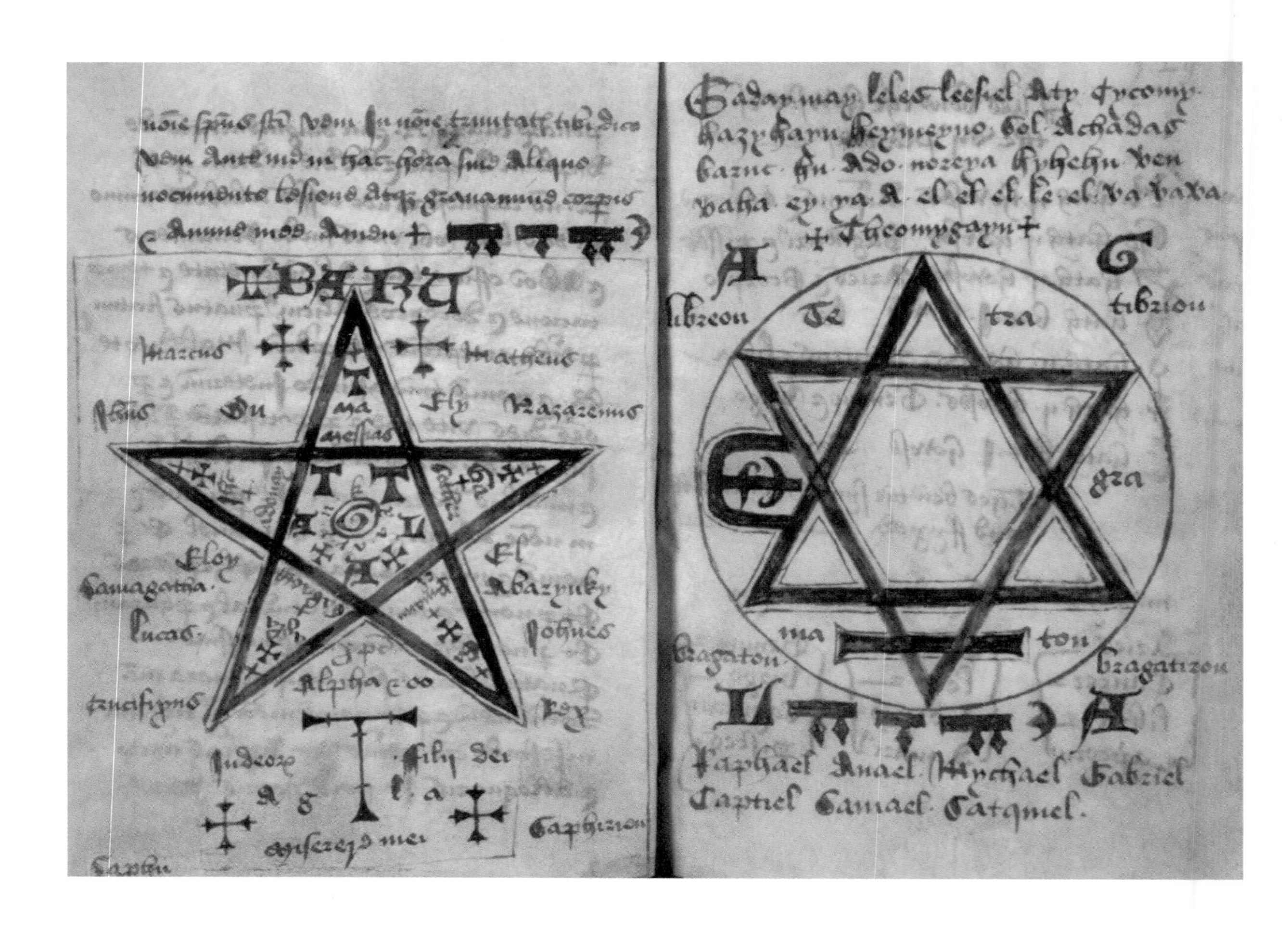

ABOVE Pages from a parchment conjuring manual in Latin and English (15th century). Much of the book is concerned with the attainment of knowledge through conjuration, including the nature of demonic powers, though there are more mundane goals as well. It includes the usual psalms and prayers, some related to the *Ars notoria* (see page 57), to support the ritual invocations and creation of talismans.

RIGHT Page from the same parchment magic book (above) in black and red ink. The image shows a lamen (magical talisman) to be created in wax to catch a thief. The ritual had to be conducted within three days of the theft, and the magician had to attend Mass at an astrologically propitious moment. The lamen includes the names of four spirits representing the compass points, and in the center is the word "Sathan" (Satan), suggesting, perhaps, a fiery punishment for the thief.

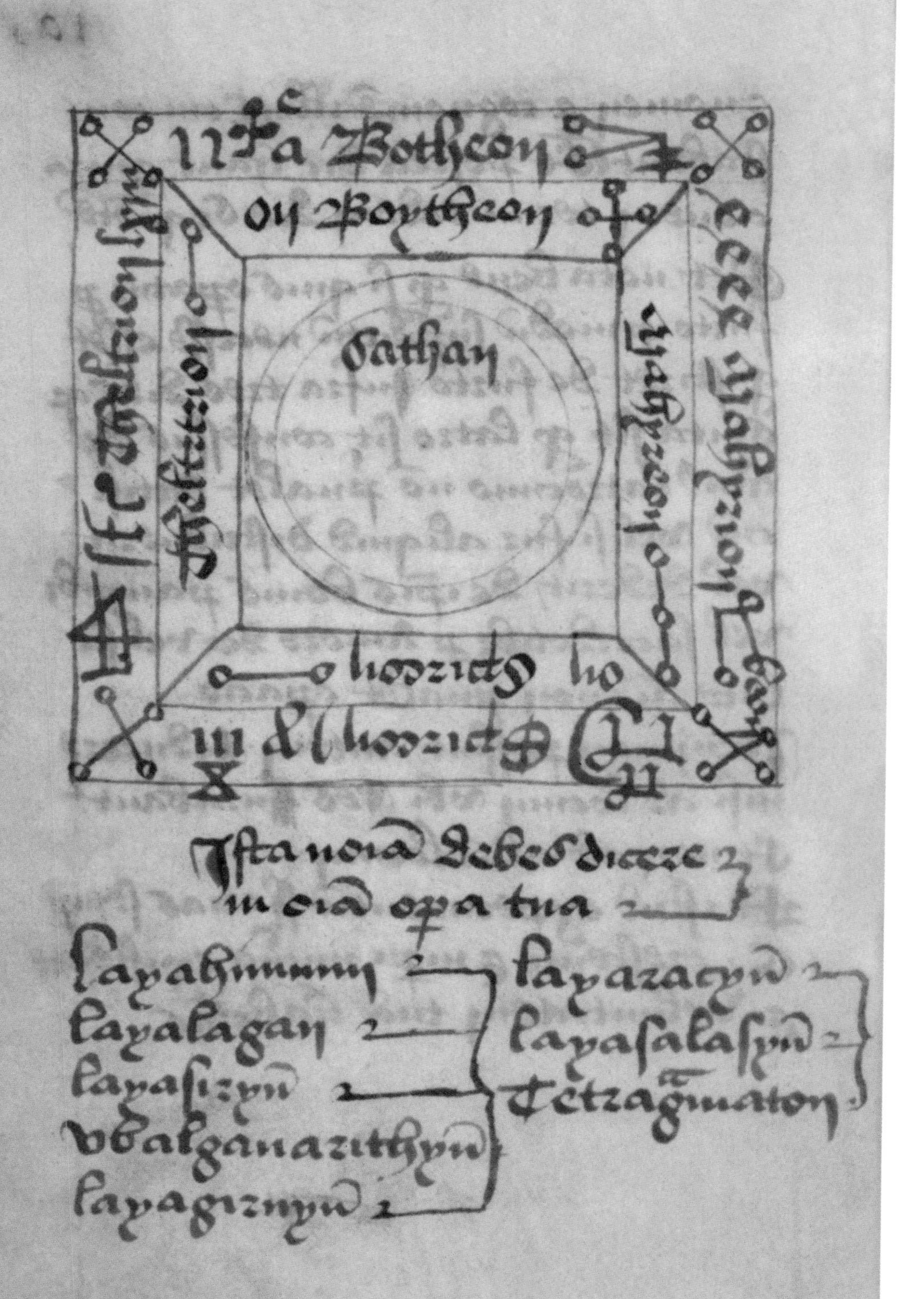
Sathan
Ista noia debes dicere
in oia opa tua

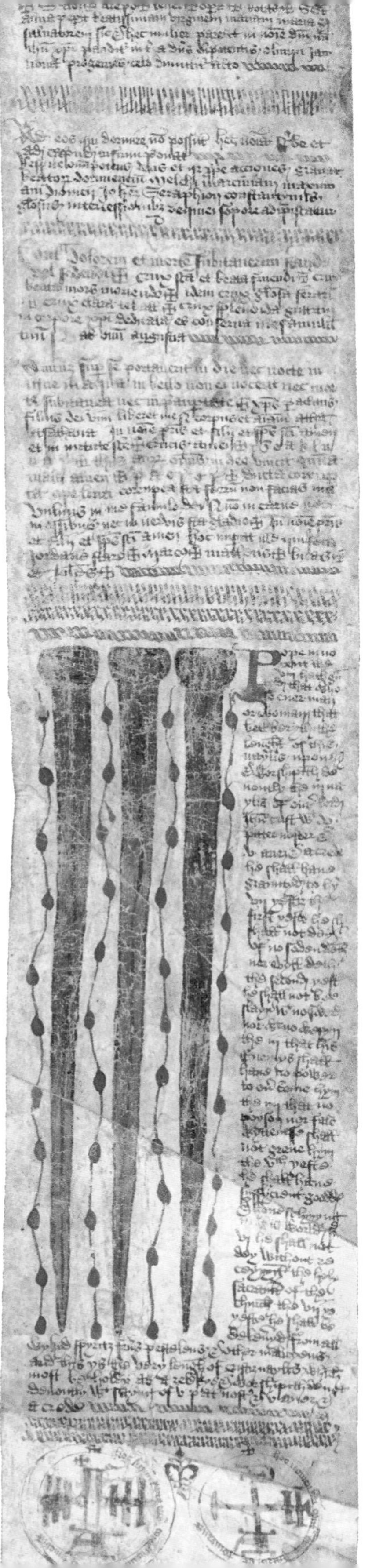

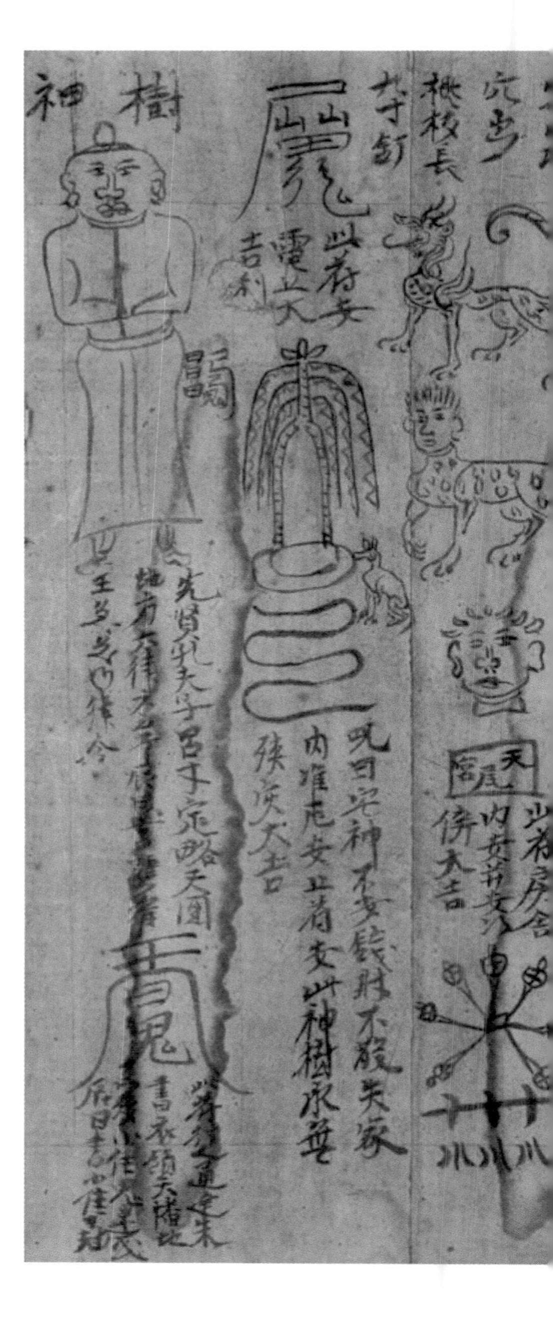

LEFT Prayer amulet roll on parchment, in English and Latin (15th century). This was made of two leaves of parchment stitched together at 48in by 3in (122cm by 8.5cm). The roll is meant to be the length of Christ's body and includes a triclavian image—a depiction of three nails, which some thought were the number used to crucify Christ. This was an old matter of theological debate, with some arguing four nails were used. As well as paternosters and prayers for protection, the roll also includes Latin invocations against insomnia, ailments, and sudden death.

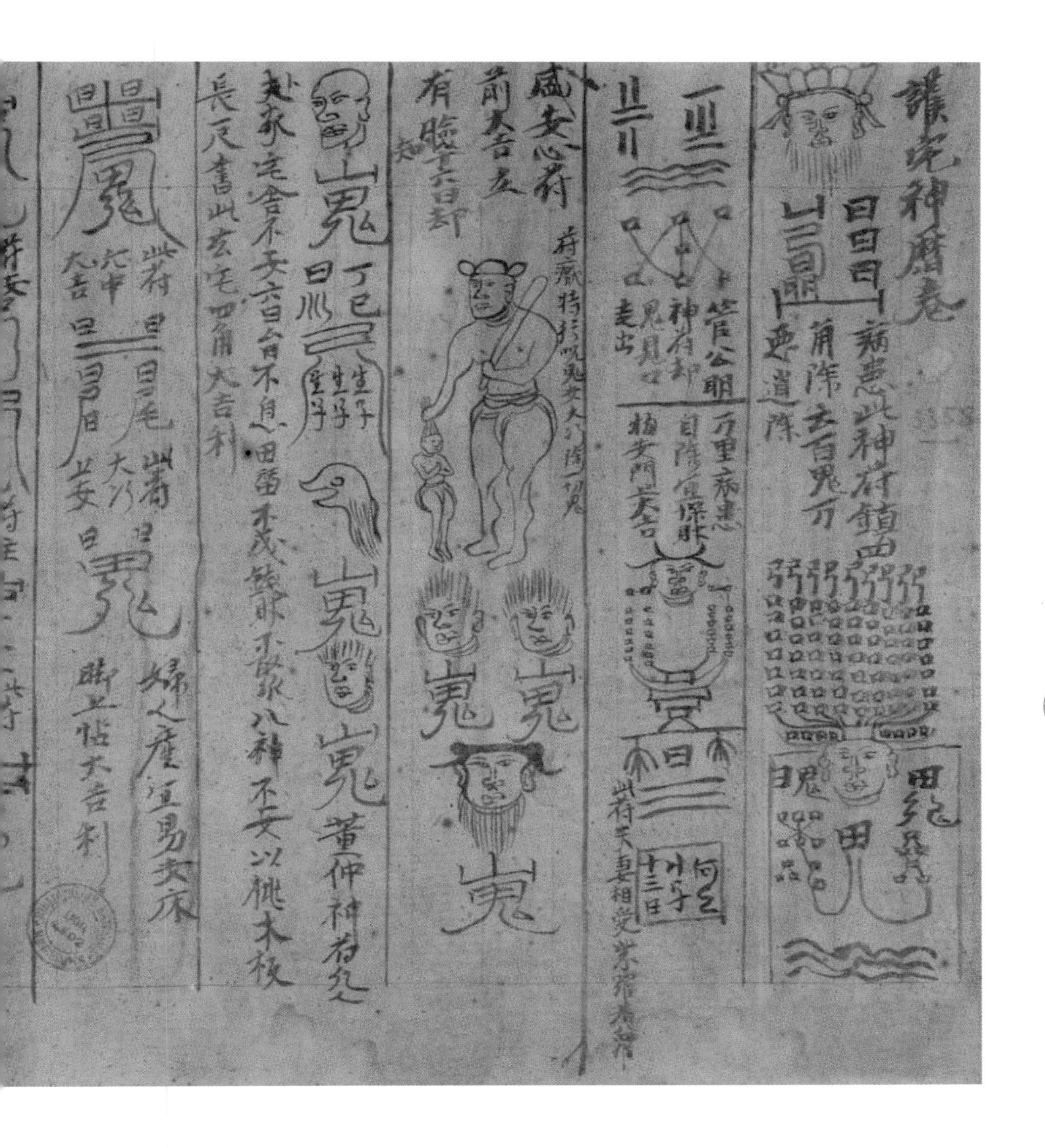

ABOVE *Huzhai Shenli juan* 護宅神歷卷, ink on paper, Dunhuang, China (10th century). This is a Chinese manual of talismans for the protection of the household. Twenty different talismans are depicted, some with instructions on how they should be fabricated, some with associated images of demons. These talismans covered a range of everyday issues including protection from nightmares and evil spirits, and prevention of infertility and family conflict.

10
Talism
Magi

CHAPTER THREE

Printing Magic

The technology of printing has its origins in China, where the use of carved wooden blocks to stamp images and text on fabric was first invented in the 2nd century CE. From the 9th century onward these blocks were being used to print on paper, too. Thousands of copies of Buddhist *sutras* were produced in this way. The earliest dated printed scroll, the *Diamond Sutra* (868 CE), was discovered in the Dunhuang caves and consists of seven woodblock-printed pages glued together to produce a scroll some 16ft (5m) in length. The manuscript tradition of *dhāranī sutras*, as well as astrological almanacs, proved highly adaptable to this new print technology, and no doubt many thousands of protective *dhāranī sutras* were produced by entrepreneurial printers in China's major urban centers from the 10th century onward.

Block-printing, using either wood or metal, also developed in the Arabic world as early as the 10th century, where the technology was used to print Islamic texts as well as protective amulets consisting of esoteric symbols and passages from the Qur'an. Only around a hundred such Arabic prints on paper and parchment exist in collections today and no early wood or tin printing block (known as a *tarsh*) can currently be located. This process eventually spread from Egypt and the Middle East to Moorish Spain, where Arabic amulets printed on rag paper appear to have circulated in Andalucía during the 13th century. Surviving examples from around this time demonstrate that the papers were folded many times and kept in small lead cases. When the Spanish Inquisition turned its attention to popular magic in the 16th century, inquisitors found that such printed and manuscript textual amulets in Arabic were still widespread and not just among the converted Muslim population.

It was over a thousand years after the invention of printing that the technology finally found its way to Central and Western Europe—and only a century after paper-making had been introduced to the region. The German goldsmith and printer Johannes Gutenberg (d. 1468) is often said to have been the inventor of moveable type, although it had already been used in limited ways in China, Korea, and elsewhere. Yet, Gutenberg certainly made significant advances in both the use of metal type and the mechanics of the printing press itself, which enabled several thousand

LEFT *Tarsh* printed amulet, colored inks on paper in *Naskh* Islamic script, North Africa or Spain (14th to 15th century). The large red script at the top reads: "There is no victor but God." Devotional texts are written within the circle, although the meaning of the letters and numbers in the outer frame of the circle are difficult to decipher.

RIGHT Woodcut of a printworks by Jost Amman (1539–1591), from Hartmann Schopper's, *Panoplia omnium illiberalium mechanicarum* (*Book of Trades*), Frankfurt (1568). By the late 16th century, the printing press had created a flourishing, international publishing industry. When it came to printing magic texts, however, publishers had to weigh up carefully the money to be made against the legal jeopardy in the age of the witch trials.

Typographus. Der Buchdrucker.

Arte mea reliquas illustro Typographus artes,
Imprimo dum varios ære micante libros.
Quæ prius aucta situ, quæ puluere plena iacebant,
Vidimus obscura nocte sepulta premi.

Hæc veterum renouo neglecta volumina Patrum
Atq; scolis curo publica facta legi.
Artem prima nouam reperisse Moguntia fertur,
Vrbs grauis, & multis ingeniosa modis.
Qua nihil vtilius videt, aut preciosius orbis,
Vix melius quicquam secla futura dabunt.
C 3 Char-

pages to be produced in just a day. Printing presses were quickly established in cities and towns across the continent, and new innovations were constantly improving both the quality, quantity, and price of books. It has been estimated that in Western Europe around 120 manuscript books were produced annually during the 6th and 7th centuries, but by the late 18th century, that figure was as high as twenty million printed books. This massive expansion of literature in print had a profound impact on access to magical knowledge.

AGE OF THE WITCH TRIALS

Seen as a symbol of the Renaissance and the rise of modern European culture, and as being instrumental in the spread of the Protestant Reformation, the Gutenberg press also coincided with the rise of the witch trials. While the early presses published a range of religious and medical books, they also printed notorious demonological texts that were key to spreading fears about a new threat to Christendom apparently posed by groups of devil-worshipping witches. Several sensational trials had occurred in central Europe where, under torture, people, mostly women, confessed to all sorts of malicious magical acts and satanic sexual relations. Evidence from these trials were documented by theologians-turned-demonologists in such works as *Formicarius* (1436–1438) by the German Dominican Johannes Nider (c. 1380–1438), which was posthumously printed in the 1470s. The object of focus for these early demonologists was not male magicians and learned necromancers with their manuscript grimoires, but poor women. As Heinrich Kramer (c. 1430–1505), the Dominican inquisitor and author of the notorious witch-hunter's manual the *Malleus Malleficarum* (1487) explained, "this kind of superstition [witchcraft] is not practised in books or by the learned but entirely by the ignorant."

Demonologists and demonological books proliferated through the 16th and early 17th centuries, and magicians and their grimoires became a more prominent concern for the religious and secular authorities. Indeed, the laws instituted against witchcraft across Europe

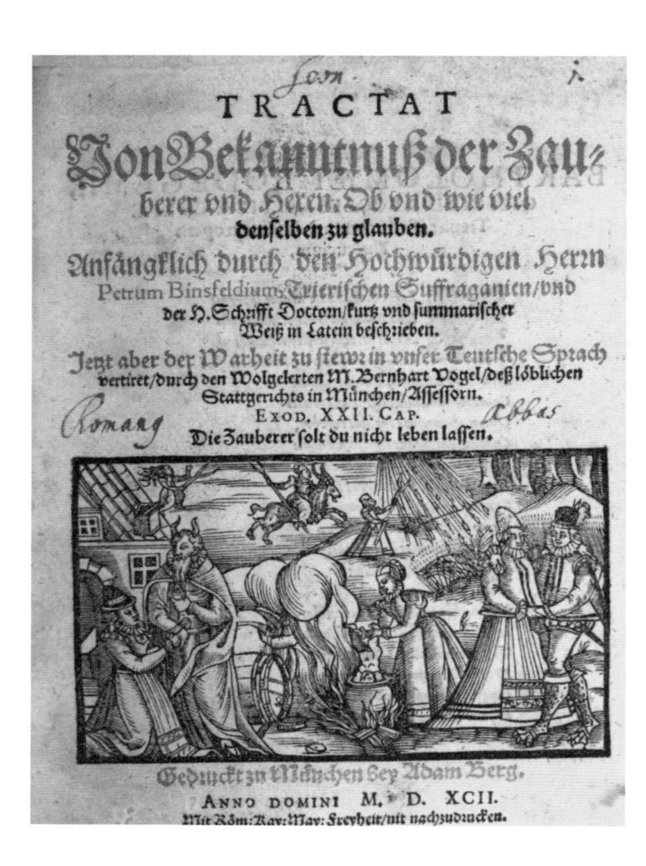
TRACTAT
Von Bekanntnuß der Zauberer vnd Hexen. Ob vnd wie viel denselben zu glauben.
Anfängklich durch den Hochwürdigen Herrn Petrum Binsfeldium, Trierischen Suffraganien/vnd der H. Schrifft Doctorn/kurtz vnd summarischer Weiß in Latein beschrieben.
Jetzt aber der Warheit zu stewr in vnser Teutsche Sprach vertiret/durch den Wolgelerten M. Bernhart Vogel/deß löblichen Stattgerichts in München/Assessorn.
EXOD. XXII. CAP.
Die Zauberer solt du nicht leben lassen.
Gedruckt zu München bey Adam Berg.
ANNO DOMINI M. D. XCII.
Mit Röm: Kay: May: Freyheit/nit nachzudrucken.

LEFT Title page of Peter Binsfeld's *Tractat von Bekanntnuß der Zauberer und Hexen* (*Treatise Concerning the Confessions of Sorcerers and Witches*), Munich (1592 edition). Binsfeld (1546–1598) was the Bishop of Trier, and believed all magic was essentially diabolic witchcraft. The woodcut shows a witch paying homage to the Devil (left), another cooking a baby (center), while another conjures a hailstorm.

also prohibited the "white" magical practices of popular magicians known as cunning-folk, who often made great play of their magic books to impress clients who came to them on matters of bewitchment, affairs of the heart, and the theft of goods. The Catholic church included named magic books in its various published *Indexes of Prohibited Books* during the 16th century. Even those who voiced open scepticism of the witch trials, such as the Protestant English gentleman Reginald Scot (1538–1599), were fierce in their criticism of popular magicians and their mysterious books.

Despite the laws, denunciations of magic, and attempts to suppress the circulation of grimoires, during the second half of the 16th century, the printing presses began to produce books of occult philosophy, astrology, conjurations, and alchemy that would continue to have an influence on the world of magical practice right through to the present day. As books were increasingly published in the vernacular and became cheaper, and literacy rates increased, so centuries of esoteric knowledge that had only been accessible to the learned elite fluent in Latin, Greek, and Arabic, slowly but surely became accessible to the many. Thanks to the printing press, the democratization of learned magic became unstoppable.

The most influential occultist of the early print age was the Renaissance German scholar Heinrich Cornelius Agrippa von Nettesheim (1486–1535), author of the *De occulta philosophica*, or *Three Books of Occult Philosophy*, first printed just before his death. Agrippa was no spirit conjuror, for he condemned the "commerces of unclean spirits made up of the rites of wicked curiosities, unlawfull charms, and deprecations" that were "abandoned and execrated by all laws. Of this kinde are those which we now adayes call Necromancers, and Witches." Yet, he was deeply immersed in natural magic, the possibility of angelic communications, the workings of the "Kabbalah of names," and the secret powers of sigils—illustrations of which were included in the *Three Books*. After his death, it is no surprise, perhaps, that Agrippa's name came to be spuriously associated with books of conjuration that he would have wholly denounced. In particular, *A Fourth Book of Occult Philosophy*, which contained spirit conjurations, was published in Latin in 1559, with another edition appearing in Basel, six years later. Further Latin editions appeared bundled with versions of the Solomonic *Ars Notoria* (see page 57). These, in turn, gave birth to vernacular manuscript variants of an "Agrippan" grimoire that circulated across Europe.

It was not only Renaissance occult philosophers who made centuries of magical wisdom available to a wider audience through print. Ironically, two of the most vocal 16th-century critics of the witch trials and of popular magic wrote books that unwittingly became major influences on manuscript grimoires throughout the next two centuries. One was the Dutch physician Johann Weyer (1515–1588), a student of Agrippa's, whose sceptical discourse *De Praestigiis Daemonum et Incantationibus ac Veneficiis* (*On the Illusions of the Demons and on Spells and Poisons*) was published in 1563. A subsequent edition included an appendix called the *Pseudomonarchia Daemonum*, which was a list of spirits and their powers—similar to those in some late medieval manuscripts—along with advice for conjuring them. Its contents derived from a spirit conjuration manuscript seen by Weyer and in printing it his aim was to show the "abominable" and ungodly nature of grimoires circulating at the time. Yet, through publication, Weyer added to an emerging print library of learned magic, along with the *Fourth Book* (which Weyer also condemned), which would disseminate the secrets of spirit conjuration further and wider than ever before.

Over in England, Weyer's *Praestigiis Daemonum* was an inspiration for Reginald Scot, who was equally sceptical of the evidence being brought against witches and was also an ardent critic of Catholic "superstition." In his book, *The Discoverie of Witchcraft* (1584), Scot not only provided an English version of the *Pseudomonarchia Daemonum* but also, for the same reasons as Weyer, produced full extracts from an English grimoire "written in fair letters of red & blacke upon parchment" that was full of conjurations and related signs, seals, and

RIGHT Title page of an edition of Giambattista della Porta's *Magia Naturalis* (*Natural Magic*), first published in Naples in 1558 (1644 edition). This hugely influential book on natural secrets by the Italian polymath della Porta (1535–1615), went through at least 20 editions and several translations. Despite being a scientific work, its title and contents gave it an undeserved reputation as a dark work of magic.

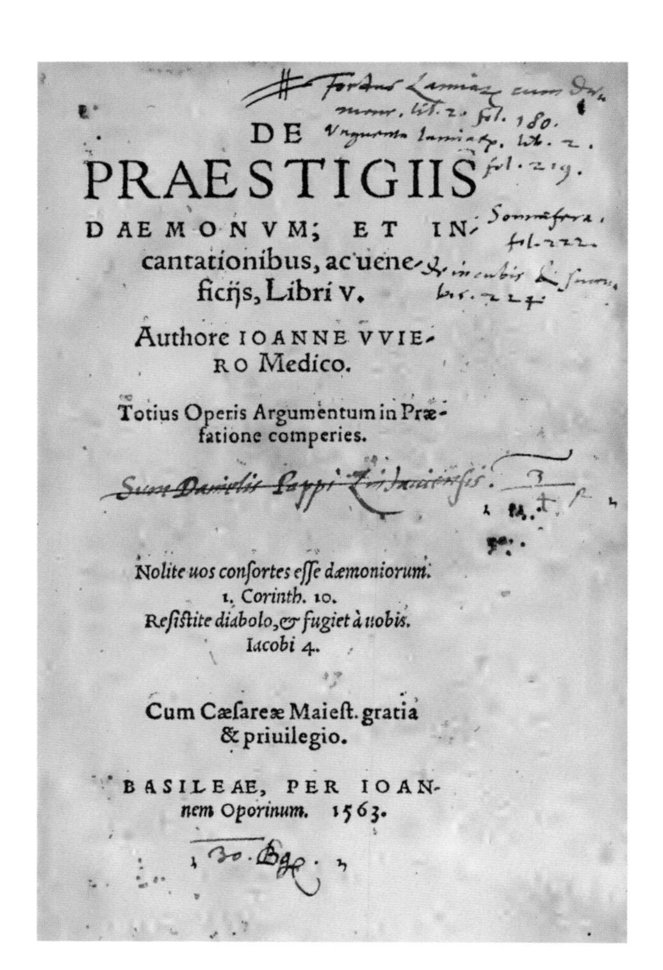
DE
PRAESTIGIIS
DAEMONVM; ET IN-
cantationibus, ac uene-
ficijs, Libri V.
Authore IOANNE VVIE-
RO Medico.
Totius Operis Argumentum in Præ-
fatione comperies.

Nolite uos consortes esse dæmoniorum.
1. Corinth. 10.
Resistite diabolo, & fugiet à uobis.
Iacobi 4.

Cum Cæsareæ Maiest. gratia
& priuilegio.
BASILEAE, PER IOAN-
nem Oporinum. 1563.

LEFT Title page of Johann Weyer's *De Praestigiis Daemonum et Incantationibus ac Venificiis* (*On the Illusions of the Demons and on Spells and Poisons)*, Basel (1563). Weyer did not deny the power and influence of the Devil on humans, but he argued that confessions by witches mostly resulted from female mental illness, and were not based on physical reality.

diagrams. Subsequently, over the next few centuries, *The Discoverie of Witchcraft* became a major sourcebook for aspiring English-speaking magicians on both sides of the Atlantic.

Early modern grimoires, whether in print or manuscript, had surprisingly little to say about witchcraft at a time when tens of thousands of (mostly) women were being prosecuted and executed as witches. This can be explained in part by the fact that magical texts continued to be copied from medieval manuscripts written at a time when trials for witchcraft were few and far between. We find more anti-witchcraft spells and charms in the compilations of practical magic written down by cunning-folk rather than university-educated magicians. As the owner of the English grimoire copied by Scot stated in his manuscript, it was written for the "maintenance of his living, for the edifying of the poore." Women also continued to be excluded from the tradition of legendary grimoire authorship, even though some cunning-women who were caught up in sensational witch trials confessed to having used magic books. Yet, 16th- and 17th-century artists who gloried in depicting the demonological accounts of orgiastic witches' sabbats also represented witches consulting books of magic, even though accounts of such "witches' books," as distinct from the books of cunning-women, were very rare in the European trials.

LEFT Alchemical engraving by Raphael Custos (1590–1664), from Stephan Michelspacher's, *Cabala: Spiegel der Kunst und Natur, in Alchymia* (Cabala: *Mirror of Art and Nature, in Alchemy*), Augsburg (1615). It concerns symbols of the states of physical matter (top), cabalistic diagrams (middle), and depictions of alchemical distillation and calcination (bottom).

ALCHEMY AND THE PRINTED IMAGE

The European print revolution also helped promote the study of alchemy as never before. Alchemy was essentially an aspect of natural magic, and medieval alchemical manuscripts provided instructions in both theory and practice, with experimental advice on how to transmute base metals to precious metals, and how to seek the mythical philosopher's stone, or elixir, that could mediate the transmutation. Muslim-Arabic literature was a big influence on the medieval European preoccupation with alchemy. But the press promoted a new Western alchemical tradition of "Hermeticism" that entwined Christian esotericism and science with the purported teachings of the legendary man-god magician of the Hellenistic world, Hermes Trismegistus. Sometime in the early centuries of the Common Era a collection of philosophical, mystical, astrological, alchemical, and religious texts attributed to Trismegistus were in circulation, and these Hermetic writings were copied and passed through the centuries by Christians and Muslim scholars in the Eastern Mediterranean. Then, in 1471, a Latin translation of Greek Hermetic works was published in Italy, which ignited a wave of scholarly interest among European astrologers, alchemists, and physicians. Among them was the Swiss occult physician Theophrastus Bombastus von Hohenheim (1493–1541), otherwise known as Paracelsus.

LEFT "Here is the figure of the spirit when it appears." Woodcut from the *Dragon Rouge* (early 19th century, see page 112). The sacrificed black hen lies at the magician's feet, within a circle scratched on the ground using a rod of cypress wood. The magician then pronounces these words three times: "ELOÏM, ESSAÏM, frugativi et appelavi."

Paracelsus is an important figure in the history of Western magic because his influential publications on medicine and alchemy, as well as the inevitable pseudo-Paracelsian books that were also printed, had a deep practical impact on the everyday lives of people in terms of the widespread application of his theories of chemical medicine. These entwined natural magic and astral magic with the empirical observation of the effects of chemical compounds on the human body. He once wrote, "the stars teach us all the arts that exist on earth; and if the stars were not active in us, and if we had been compelled to discover everything in ourselves, no art would ever have come into being."

The 17th century has often been described as the golden age of alchemy as it was deeply entangled with the development of modern chemistry in the West. Many leading scientific figures of the time, such as Isaac Newton (1643–1727), were certainly interested in, or engaged in, alchemical investigations. Yet, the idea of a golden age is more convincingly argued in terms of alchemical aesthetics. Early-modern printing could not match the vibrant colors and use of gold leaf in some late medieval and Renaissance alchemical manuscripts, but, during the 17th century, developments in print copper-plate engraving, such as etching with acid, increasingly usurped woodcut images in high-quality publications. The intricate fine lines now made possible produced ornamental, alchemical, and cosmological art of extraordinary detail: it was, in a sense, an alchemy in its own right.

Despite the advances in the science and art of metal-plate engraving, the woodcut magical image was far from dead. It was, after all, much cheaper to produce and print. Woodcuts were the standard artistic form for illustrating British witch-trial pamphlets, divination manuals, and astrological almanacs, for example, while in 18th- and early 19th-century France, they were integral to the success of a new genre of cheap, mass-produced grimoires that lit up the magical imaginations of the poor and illiterate. There were a range of titles that referenced the medieval magical tradition, and they usually claimed to have spurious publication dates in the 16th and 17th centuries. The most notorious of these was the *Grand Grimoire*, or *Dragon Rouge*, because it contained an elaborate conjuration to call up a devil at a crossroads after sacrificing a black hen at midnight: "His head will resemble that of a dog with donkey's ears, with two horns above them; his legs and his feet will be like those of a cow. He will ask for your orders."

By the end of the century, similar, cheap print productions began to appear in neighboring German territories, though very few survive today—perhaps due to their confiscation by the authorities. One of the first to be printed was entitled *Trinum perfectum magiae albae et nigrae* (*Perfect trinity of black and white magic*), which contained instructions for commanding spirits. It was purportedly written by a mysterious rabbi called Tabella Rabellina, whose name we find mentioned in other German manuscript and print grimoires of the period. His supposed work was cited, for instance, as a sourcebook in another cheap print magic book, *D. Fausts Original Geister Commando* (*Dr. Faust's Original Spirit Command*). This was one of a genre of popular Faustian *Höllenzwang*, or "Coercion of Hell," grimoires that supposedly contained the magical secrets of Dr. Faust. Some thirty different *Höllenzwang* texts are recorded as having been published up to 1930, though several are now lost. We will discover more about these Faustian grimoires in the next chapter.

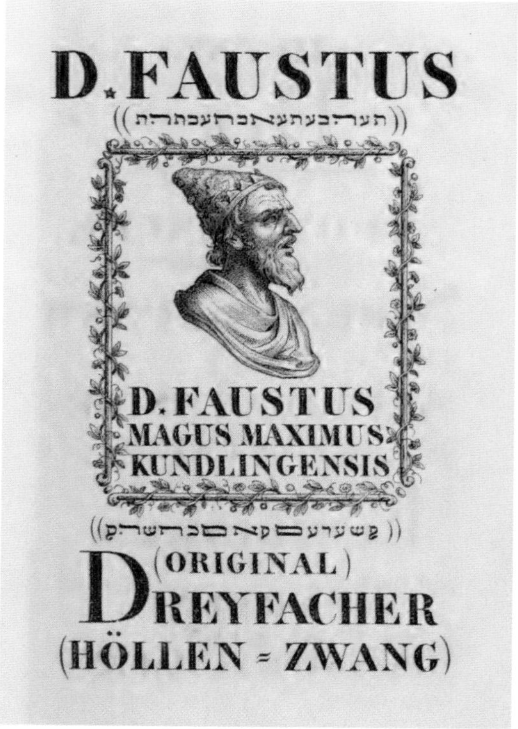

RIGHT Title page of *D. Faustus Magus Maximus Kundlingensis*, a cheaply produced printed grimoire, Germany (probably late 18th century). One of several such printed Dr. Faust *Höllenzwang* or "Coercion of Hell" magic books, which were based on spirit conjuration manuscripts that began circulating across German territories and Denmark from the late 17th century.

RIGHT Amulet scroll with polychrome block print, Egypt or Iran (10th to 11th century). With its use of green, red, and black inks, this rare *tarsh* print (see page 90) contains passages from the Qur'an, including the oft-used passage in Islamic amulets: "Assistance from God and a speedy victory." It also lists the 99 holy names of Allah. While used for general protection, such early printed amulets may have been produced primarily for warriors and for military success, like the wearing of a *taweez* (Qur'anic passages contained in small pouches or containers) by Islamic soldiers in later centuries.

LEFT Talismanic printed scroll, probably from Egypt (11th century). The angular script used is *kufic*, which takes its name from the city of Kufa, Iraq. It was one of the earliest forms of Arabic calligraphy and the preferred style for copying the Qur'an. The six-pointed star was a form of Solomon's seal, depicted widely in Islamic art, architecture, and magic.

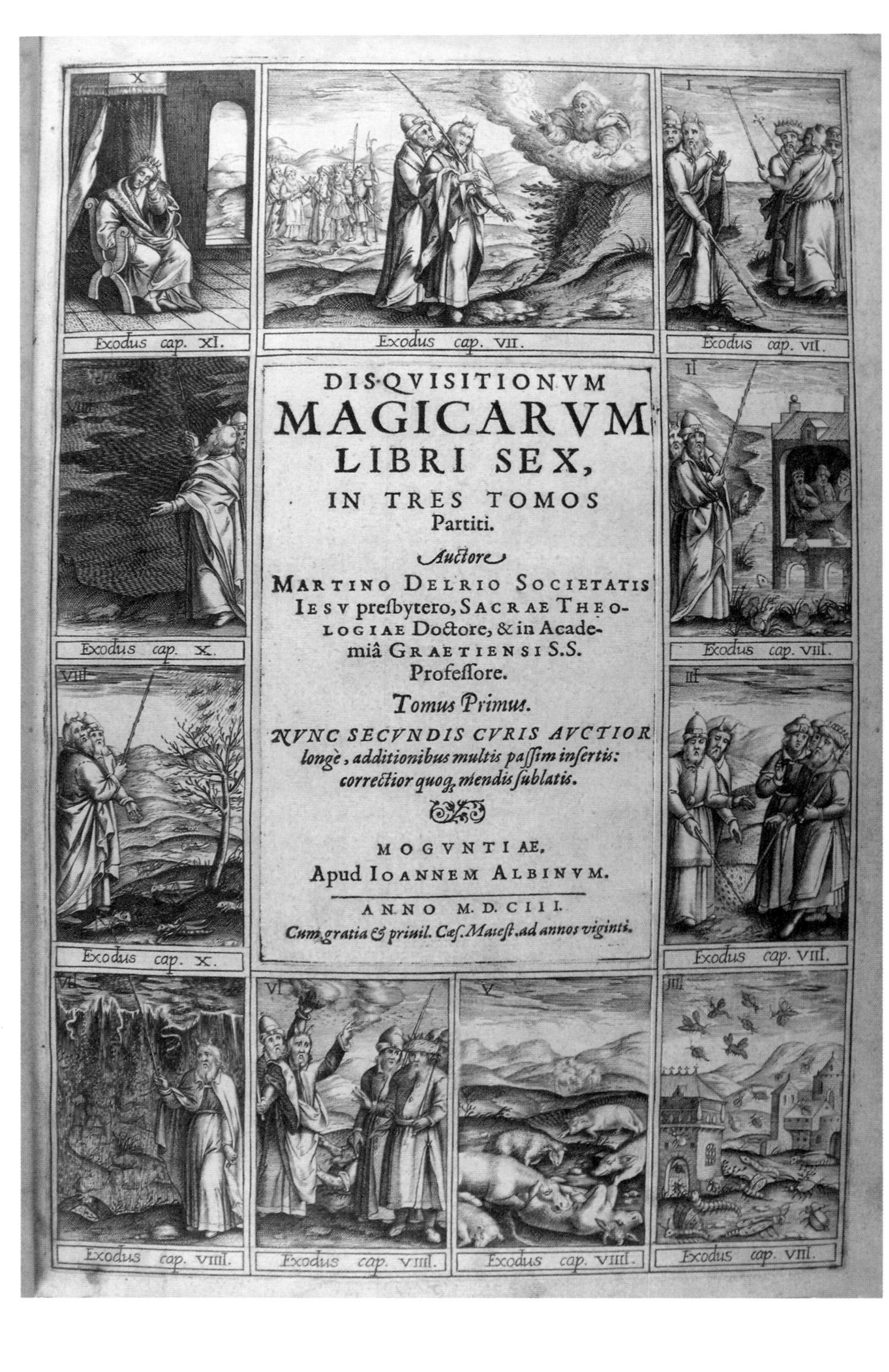
Exodus cap. XI.
Exodus cap. VII.
Exodus cap. VII.

DISQVISITIONVM
MAGICARVM
LIBRI SEX,
IN TRES TOMOS
Partiti.
Auctore
MARTINO DELRIO SOCIETATIS
IESV presbytero, SACRAE THEOLOGIAE Doctore, & in Academiâ GRAETIENSI S.S. Professore.
Tomus Primus.
NVNC SECVNDIS CVRIS AVCTIOR longè, additionibus multis passim insertis: correctior quoq; mendis sublatis.

MOGVNTIAE.
Apud IOANNEM ALBINVM.
ANNO M.D.CIII.
Cum gratia & priuil. Cæs. Maiest. ad annos viginti.

Exodus cap. X.
Exodus cap. VIII.
Exodus cap. X.
Exodus cap. VIII.
Exodus cap. VIIII.
Exodus cap. VIIII.
Exodus cap. VIIII.
Exodus cap. VIII.

LEFT Martin Del Rio's *Disquisitionum Magicarum Libre Sex* (*Magical Investigations in Six Books*). First printed in three parts between 1599 and 1600, this was one of the most influential Catholic demonological books in the era of the witch trials, and it was reprinted numerous times. It was also the most widely distributed demonological work in Spanish colonial America. Del Rio (1551–1608) was born in Antwerp (in the Spanish Netherlands at the time), and he became a highly educated politician and Jesuit. For Del Rio, magical practices were usually inspired by the Devil, and hence were heresy. While he accepted some of the intellectual ideas of such occult philosophers as Agrippa and Paracelsus, he warned against the uneducated dabbling in such matters for fear of being corrupted by Satan.

ABOVE David Teniers the Younger, *Witches' Initiation*, oil on wood (1647–1649). The Flemish Teniers (1610–1690) was a successful artist known for painting the commonplace, such as tavern scenes, as well as alchemists' laboratories. The *Witches' Initiation* depicts an older woman instructing the younger one in the ways of witchcraft, as contained in books of magic. The impeccable clothing of both figures portrays them as comfortably-off Dutch women, far from the hag-like image of witches in other paintings of the period. They are surrounded by magical objects and devilish spirits, and in the background a witch is smearing a flying ointment on another naked witch, who is ready to fly up the chimney.

Dr. Faust

Thanks to demonologists and witch-trial pamphlets, we know a great deal about the identities and activities of popular magicians, who serviced the many fears and concerns of common people—and, more to the point, how they used print and manuscript grimoire to help their clients. Yet, the power of print also created a new generation of legendary magicians, who joined the pantheon of Solomon, Honorius, and the like. Foremost among these was the notorious Dr. Faust. The earliest description of the historical Faustus appears in a letter written by abbot and occult philosopher, Johannes Trithemius (1462–1516). According to Trithemius, the real Dr. Faust was a boastful traveling showman who called himself the "prince of necromancers:"

> "He said in the presence of many people, that he had acquired such knowledge of all wisdom and such a memory, that if all the books of Plato and Aristotle, together with their whole philosophy, had totally passed from the memory of man, he himself, through his own genius, like another Hebrew Ezra, would be able to restore them all with increased beauty...that he himself could do all the things which Christ had done."

Later, decades after his death, Faust was transformed into a European legendary magician thanks to the printing press. The first print versions of the fictional Faust appeared in German in the late 16th century. The stories recount how

LEFT Colored print entitled "The Devil and Dr. Faustus," (c. 1825). The fascination with the Faust story—fueled by popular literature and art of this kind over the centuries—influenced impressionable young men to seek and emulate the notorious magician, using the grimoires attributed to him to conjure up the Devil.

RIGHT Color illustration from *The Remarkable Life of Dr. Faustus*, London (c. 1823). Selling for six pence, this edition of the story, from prominent children's literature publishers Dean & Munday, was an abridged version of an early-modern published Faust history. Note the representation of Mephistopheles in a monk's cowl rather than as an overtly demonic presence (see also page 153).

Faust made a pact with Mephistopheles, an agent of the Devil, to enhance his magical knowledge and acquire new powers over others—until such time as the pact expired and Satan took him, body and soul. Editions of this legend were printed in Danish, French, and English by 1600, and over the ensuing centuries, the Faust legend inspired numerous artists and writers, including the German poet and playwright Johann Wolfgang von Goethe (1749–1832). The story was reprinted many times in popular literature across Europe well into the 19th century, as a warning to those attracted by the promise of diabolic powers.

LEFT Jacob Cornelisz van Oostsanen, *Saul and the Witch of Endor*, The Netherlands (1526). The Old Testament account in *1 Samuel 28* of how the "Witch of Endor" apparently raised the spirit of the prophet Samuel was a common topic for late medieval and early modern artists, and was depicted in woodcuts, engravings, and paintings. This rendering of the tale by the Amsterdam artist Van Oostsanen (c. 1475–1533) was painted decades before witchcraft became a major concern in the city. His inspiration was the less contemporary, more classical Roman notion of sorceresses. It is, however, one of the earliest paintings where the visual link is made between witches and books of magic.

ABOVE AND RIGHT *Above.* Depiction of the execution of Verena Trostin, Barbara Meyerin, and Anna Langin for witchcraft, colored pen ink drawing, Bremgarten (1574). *Right.* Hand-colorized print broadsheet reporting how devils caused terrible storms in Ghent, Flanders, in 1586. Both images are taken from the so-called "*Wonder Book*" of the Swiss pastor Johann Jakob Wick (c. 1522–1588). Wick's *Wonder Book* was a collection of hundreds of broadsheets and pen ink drawings that reported strange and terrifying events in his time. We do not know if he commissioned drawings and the colorization, or merely collected images in circulation. Together, they add up to a comprehensive artistic depiction of the everyday supernatural in the 16th century.

OVERLEAF Pen ink drawing of a werewolf attack and subsequent execution, from Johann Jakob Wick's *Wonder Book* (c. 1580). It depicts a man from Geneva being tortured with red hot pincers, and destined to be executed for the killing of 16 children in the city, while in his wolf state. The drawing shows clear influences of earlier, widely distributed images of werewolf atrocities arising from the prosecution of suspected werewolves in Switzerland and France during the 16th century.

Newe Zeytung auß Ghendt/ in Flandern.

Wie es da selbst ein gantz greülichs vnd Erschröcklichs/ vngewitter entstanden des gleichen vormals nie erhört wordn ist/ geschehen Anno 1586. den 18. Augusty.

SEhr kleglich in der zeit gemelt/
Da hat sich der Himmel gestelt.
Schröcklich/ die Winde theten auch webē
Zu dem sach man in lüften/ Schweben/
Vnter aim vngewiter groß.
Sichtbarlichen den Satan Bloß/
Auch wie er sich thete befleisen.
Villerley schaden zu beweisen/
Des manichen Mann hat geschwirtzt.
Er hat vil Heüser vmb gestürtzt/
In der Stat Ghendt/ Nache vnd feren.
Nach sollichem ließ Er sich hörren/
In aim Castel/ da griff Er ann.
Gar schröcklich ainen Edelman/
Auß seinem Leib/ vil stückh Er Riß.
Wunderlich ists/ doch gleichwol gwiß/
Das er ain klaines Heüßlein fein.
Deß aines Schuchmachers thet sein/
Hin weckh trueg/ mit sein macht entreckht.
Ein Alte Fraw darmit erschreckt/
Das man sie hat fir Thot vmbzogen.
Ferner ist der Böß Geist geflogen/
Mit lautem Getümel ohn trawren/
Auff S. Peters Kirch von den Maurē.
Stücker Riß/ vnd Rab geworffen hat/
Er fandt auch beim Waser am gstat
Einen guten frumen man Alt.
Den nam er/ vngitiger gstalt/
Bein schultern/ fürt jn in die lüfft.
Groses Hertzenleidt err jm stifft/
Gleich woll thet er in jn den dingen.
Ohn schaden widerumb herbringen/
Jttem er thet sich auch erregen.
Mit sehr grawsamen Donerschlägen/
Auch Hellen Blitzen vngehewr.
Die scheinen thetten wie das Fewr/
Zu Gott Ruften die lieben Frumen.
Sie meintē der Jüngst Tag würt kumē/
Das wert vber drey gantze Stundt.
Vnd ist offtermals thue ich Kundt/
Der Himmel Finster worden Schnel.
Als seis Nacht/ vnd dan wider Hell/
Sach gleich als wan es Brinnen thet.
Weyter fur der Teüffel verstebt/
Auff ein Bleich/ da vil Böß verricht/
Da Nun: die Grausamlich Geschicht.
Durch Gots Erbarmung nam sein End
Da gieng das Volck zu samen bhendt.
Befahlen sich dem lieben Gott/
Zu Mechell/ vmb die zeit mit Not.
Schlugs Stäin die sahen mit beschwerd
Als ob es Fewrig Khuglen weren/
Das die Menschen auß dem Felde liefen.
Hailliger Geist zu dier wir Rieffen/
Gieb vñ Gnad/ das wir doch Recht lebē.
Auff das vns auch nit die vmb geben/
Ein solche gfahr/ Greülicher Gstalt.
Bhüt vns auch vor des Satans gwalt/
Das er vns nit zu füge schaden.
Aber: die Welt ist so beladen/
Mit Frewel thut auch was jr gefelt.
Obs Gott schon für ein Grewel helt/
So khert sie Sich doch daran nicht.
Biß er die Verwuckt Welt entwicht/
Durch Christū am Jüngsten Tag Richt.

FINIS.

Getruckt zu Augspurg/ bey Hanns Schultes/ Brieffmaler vnd Formschneyder vnder dem Eisen berg.

HENRICI
CORNELII AGRIP-
PÆ LIBER QVARTVS
DE OCCVLTA PHI-
loſophia, ſeu de Ceri-
monijs Ma-
gicis.
*

Cui acceſſerunt, Elementa Magica Petri de Abano, Philoſophi.

Marpurgi Anno Domini.

1559.

ABOVE *The Fourth Book of Occult Philosophy*, Marburg (1559). The title grandly claims this was written by the renowned occult philosopher, Heinrich Cornelius Agrippa (1486–1535), but Agrippa would have denounced its contents. The first part of the book contains an account of geomancy and a general discourse on the characters and appearance of good and bad spirits, as well as the means by which to conjure them. As the author of the *Fourth Book* notes, though, the budding magician needed more practical guidance and hence the inclusion of *The Heptameron: Or, Magical Elements*, spuriously attributed to the astrologer-physician Peter de Abano (c. 1250–1316).

86 Elementa Magica

Figura Circuli pro prima hora Diei Dominicæ, veris tempore.

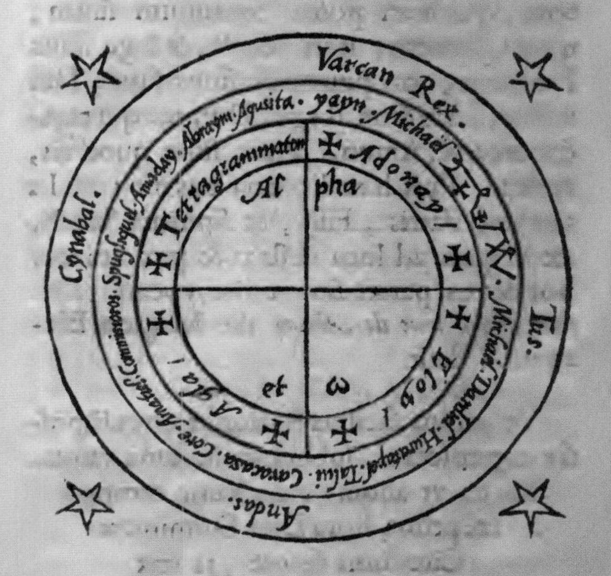

SVPEREST nunc vt Hebdomadam exploremus, ſingulóſque illius Dies, & Spiritus qui illis præſunt: & primò de die Dominico.

Conſyd

Petri de Abano. 87

Conſyderationes Diei Dominicæ.

Angelus Diei Dominicæ, & Sigillum eius, Planeta eius, & ſignum Planetæ, & nomen quarti Cœli.

Michaël. ☉. ♌.

Machen.

Angeli diei Dominicæ.

Michaël.
Dardiel.
Huratapel.

Angeli Aëris regnantes die Dominico.

Varcan Rex.

Miniſtri eius.

Tus.
Andas,
Cynabal.

Ventus cui ſubſunt Angeli Aëris ſupradicti.

Boreas.

F 4 Angeli

ABOVE *The Heptameron: Or, Magical Elements*, included in *The Fourth Book of Occult Philosophy*. The circle bears holy names, and was to be made for spirit conjurations in spring-time during the first hour of the Lord's day (Sunday). The opposite page lists the angels that could be conjured on that day, including the well-known archangel Michael (with his sign), but also lesser-known spirits, such as Varcan, the king or angel of the air, thought to have dominion over the Sun, and his angelic ministers, Cynabal, Andas, and Tus.

The Dragon Rouge

In the mid-18th century, a sensational and cheap book of ritual magic was published illicitly in France, which set out how to conjure up one of the Devil's ministers, Lucifugé Rofocale. Entitled the *Grand Grimoire*, the book provided the instructions that were missing from the stories of Faust's great conjuration of Mephistopheles. Legends grew that one could call up the Devil by merely touching the book. Yet, the *Grand Grimoire* gained even greater notoriety in French society when a variation on the text was published widely in multiple editions under the name of the *Dragon Rouge*, or *Red Dragon*. Distributed by humble *colporteurs*, or traveling salespersons, and on sale in urban bookshops, it sold in its many thousands across the country. In 1861, one commentator noted it was on open view in the windows of Parisian booksellers, "to the great scandal of those who think we are progressing."

The *Dragon Rouge* was best known for the inclusion of the *Poule Noire*, or *Black Hen*, a notorious ritual to conjure the Devil for the purposes of finding treasure. The instructions were prefaced with a story on the discovery of this ancient knowledge: "the famous secret of the Black hen, a secret without which we cannot count on

LEFT Woodcut of the Devil from the *Dragon Rouge* (early 19th century). The accompanying text reads: "Follow me, and take the treasure that I am going to reveal to you." The representation of the Devil with cloven hooves and tail was common in popular literature and in folk tales.

BELOW A woodcut copy of an emblem from the influential 16th-century alchemical book, *De Alchimia* (see pages 120 and 121), which the printers included in the *Dragon Rouge* purely for occult decoration. Its accompanying cryptic description in *De Alchimia* is: "Here Sol plainly dies again. And is drowned with the Mercury of the Philosophers."

RIGHT The frontispiece emblem of the *Dragon Rouge* (early 19th century). The title page bears the statement: "Approved by Astaroth" (one of the Devil's arch-demons) and bears his sign, or character, in red ink. We know from French court cases that people really did carry out the conjuration of such devils, as instructed in the *Dragon Rouge*, in order to find treasure.

the success of any *cabale*, was lost for a long time; after much research, we managed to find it, and the tests that we have made assure us that it was indeed the one for which we were looking."

As to the ritual:

> "Take a black hen which has never laid an egg and which no cockerel has approached; make sure on taking it that it does not cry out, and for this you will go at eleven o'clock in the evening when it is asleep, take it by the neck, making sure that it is prevented from squawking; go to a highroad, and come to a crossroads, and there, as midnight strikes, make a circle with a rod of cypress wood, place yourself in the middle of it and cut the body of the hen in two while pronouncing these words three times: ELOÏM, ESSAÏM, *frugativi et appelavi*. Turn your face to the east, kneel down and utter the prayer on page 85."

The prayer mentioned was reproduced in Latin and was entitled *Citatio Prædictorum Spirituum*. This prayer was first printed in Weyer's *Pseudomonarchia daemonum* (see page 93), and Reginald Scot included an English version in his *Discoverie* (see page 93).

PARACELSUS
Of the
Supreme MYSTERIES
OF
NATURE.

Of { The Spirits of the Planets.
Occult Philoſophy.

The Magical, Sympathetical, and Antipathetical CURE of Wounds and Diſeaſes.

The Myſteries of the twelve SIGNS of the ZODIACK.

Engliſhed by *R. Turner*,
φιλομαθής.

Jacob Neilson

London, Printed by *J. C.* for *N. Brook* and *J. Hariſon*; and are to be ſold at their ſhops at the Angel in Cornhil, and the holy Lamb neer the Eaſt-end of *Pauls*. 1656.

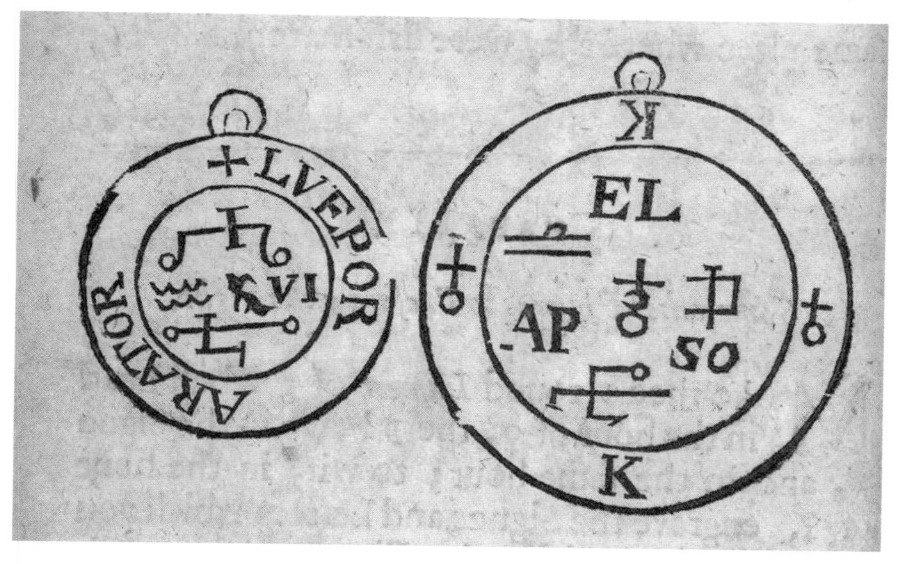

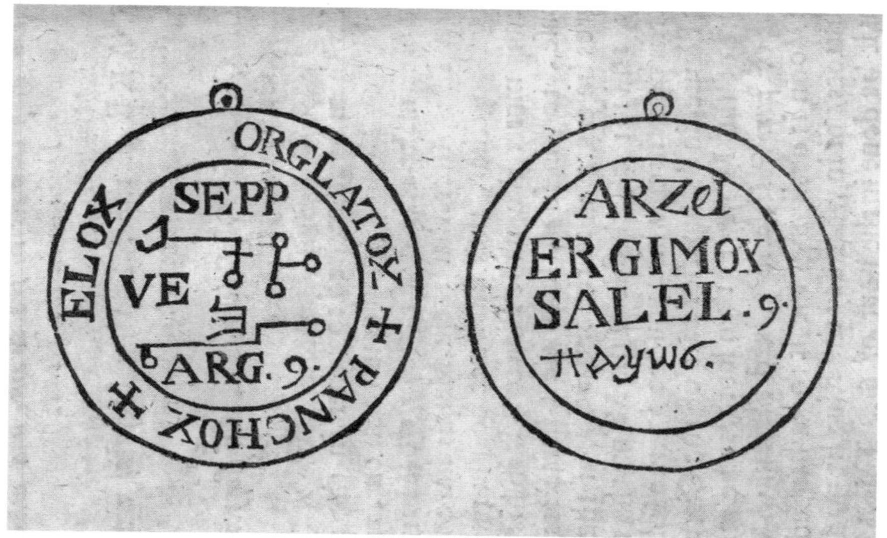

LEFT *Paracelsus of the Supreme Mysteries of Nature*, London (1656). While the author is stated as Paracelsus (d. 1541), and some of its contents on astral medicine fitted his medical theories, scholars have cast doubt that this book was written by the Swiss occult physician. The text was originally published in Latin in the 1570s, and proved popular over the ensuing years; the Paracelsus attribution was part of its success. It contained a mixture of advice on alchemy, the interpretation of dreams and visions, and the creation of healing astral lamens to be worn for protection.

ABOVE *Paracelsus of the Supreme Mysteries of Nature*, London (1656). *Above.* These two sigils, or signs, were meant to preserve eyesight in old age and keep it "as perfect as it was in youth." The designs were to be made on both a round lead and copper lamen at a propitious astrological moment. Then they were "to be conjoyned together so, that the Characters and Signes may mutually touch one another; then close them fast with Wax, that they receive no moisture, and sew them up in a piece of Silk, and hang it about the Neck of the Patient." *Below.* Sigils for diseases of the brain. These were to be engraved on lamens made of a mixture of gold, silver, copper, and tin at the time of the new Moon.

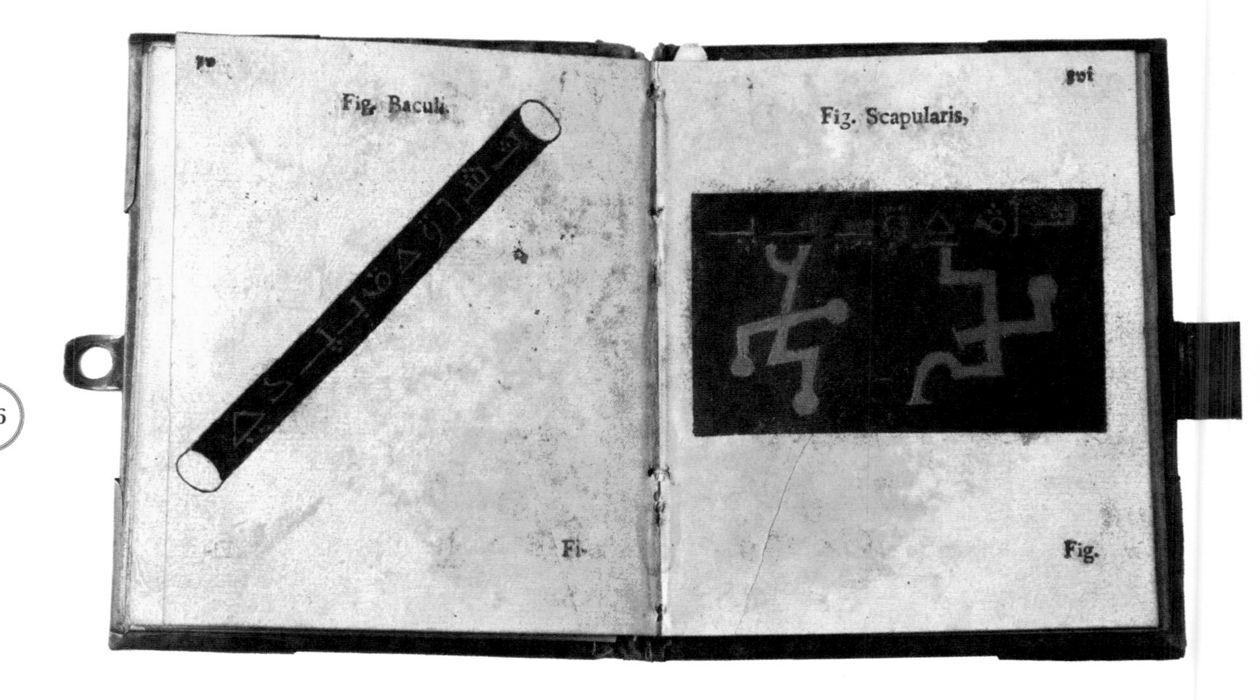

ABOVE Pages from the *Compendium magiae innaturalis nigrae* (*Compendium of Unnatural Black Magic*), written in Latin (late 16th century). This rare print book of magic, with some black and red hand-colored images, consists of only 24 pages. The preface claims it was translated from an Arabic manuscript, and its contents concern the conjuration of an Arabic-sounding spirit called Almuchabzar. The image on the left, here, accompanies instructions to cut a hazel rod, three feet in length, and the woodcut on the right relates to the creation of a breastplate from virgin parchment. Both were to be inscribed with Arabic-looking symbols using dove's blood. The magician was also instructed to create a crown and magic circle from more virgin parchment. The sigils or characters of spirits were to be written down in raven's blood.

RIGHT Frontispiece woodcut illustration from the *Compendium magiae innaturalis nigrae* (c.1570s). Its significance to the conjuration is not clear from the contents of the grimoire, but it is likely that this image represents Venus holding a flaming sword and a key. It is possible the printer borrowed a woodcut block from a previous publication for effect, as other images of Venus holding a key are known. The flaming sword was a common theme in classical mythology.

The discouerie of witchcraft,

Wherein the lewde dealing of witches *and witchmongers is notablie detected, the* knauerie of coniurors, the impietie of inchan- *tors, the follie of soothsaiers, the impudent fals-* hood of cousenors, the infidelitie of atheists, *the pestilent practises of Pythonists, the* curiositie of figurecasters, the va- *nitie of dreamers, the begger-* lie art of Alcu- mystrie,

The abhomination of idolatrie, the hor- *rible art of poisoning, the vertue and power of* naturall magike, and all the conueiances *of Legierdemaine and iuggling are deciphered:* and many other things opened, which *haue long lien hidden, howbeit* verie necessarie to be knowne.

Heerevnto is added a treatise vpon the *nature and substance of spirits and diuels,* &c: all latelie written *by Reginald Scot* Esquire.

1. Iohn. 4, 1.

Beleeue not euerie spirit, but trie the spirits, whether they are of God; for manie false prophets are gone out into the world, &c.

1584

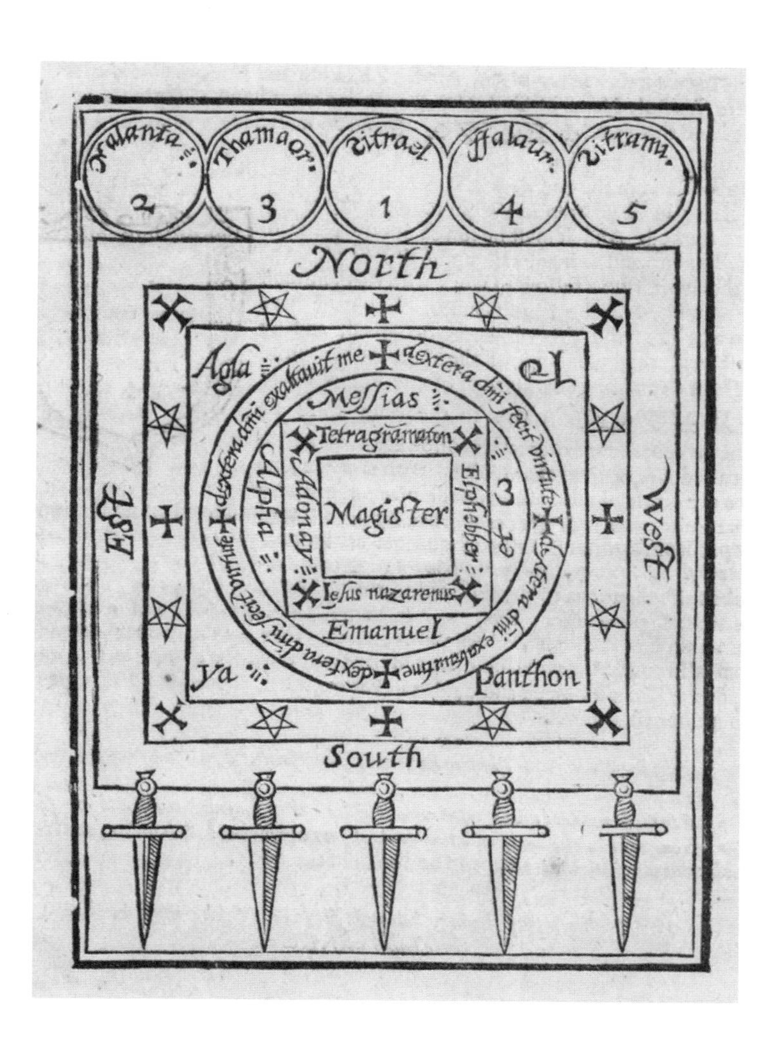

LEFT Title page from *The Discoverie of Witchcraft* by Reginald Scot, London (1584). Despite being a harsh critique of magic, Scot's book was rich in spells, conjurations, exorcisms, and talismans culled mostly from numerous continental demonological treatises. He also included the manuscript grimoire of an English cunning-man, which was entitled *Secretum secretorum* (*The secret of secrets*). It contained conjurations and ceremonies akin to those circulating in manuscripts attributed to Solomon and Honorius. Scot reproduced all of this material in order to show the vanity and foolishness of magic, but, ironically, *The Discoverie* became a hugely influential print grimoire. We find its contents copied in the manuscript magic books and charms of British magicians through the centuries.

ABOVE Image from *The Discoverie of Witchcraft* by Reginald Scot, London (1584). This elaborate diagram of circles, squares, symbols, magical words, and spirit names accompanied a conjuration to enclose a spirit in a crystal. The diagram was to be marked out on the ground according to a ritual order. The magician had to wear new clothes, fast, and pray for two days before performing the ritual. He also needed five swords, as represented at the bottom of the image. The names written at the top were described as the five infernal spirit kings. After several days of exhausting conjurations, the five kings would appear from the north with a marvellous retinue. They would come to a stop at the circle, kneel down, and declare: "Maister, command us what thou wilt."

DE ALCHIMIA

OPVSCVLA

COMPLVRA VETERVM PHI-
loſophorum, quorum catalogum ſe-
quens pagella indicabit.

Cum gratia & Priuilegio Cæſareo.

ROSARIVM

Wyr sindt der metall anfang vnd erste natur/
Die kunst macht durch vns die höchste tinctur.
Keyn brunn noch wasser ist meyn gleych/
Ich mach gesund arm vnd reych.
Vnd bin doch jtzund gyftig vnd dötlich.

Succus

PHILOSOPHORVM.

& sicca soluantur, calcinentur, siue sublimentur se cundum quod viderit, & melius iudicatur secun= dum sanum sensum operantis.

Ich bin der war grün vnnd guldisch Löwe ohn sor= (gen/
Inn mir steckt alle heimlichkeyt der Philosophen verborgen.

Y iij

LEFT Title page from *De Alchimia: Opuscula complura veterum philosophorum (On Alchemy: Several Works of the Ancient Philosophers)*, Frankfurt (1550). First printed in 1541, this compilation of alchemical writings includes manuscript treatises spuriously attributed to various early medieval figures, including the 7th-century Arab prince Khālid ibn Yazīd, and what is known as the "Pseudo-Geber," attributed to the legendary 8th-century scientist Jābir ibn Hayyān. This version of *De Alchimia* from 1550 was influential, in part, due to the 20 woodcut emblems (symbolic coded images) that take the reader through an alchemical spiritual journey.

ABOVE Woodcut emblems from the *Rosarium Philosophorum*, part of *De Alchimia* (1550). The image on the left represents the fountain of life, and the associated epigram explains: "We are the beginning and first nature of metals, Art by us maketh the chief tincture. There is no fountain nor water found like unto me. I heal and help both the rich and the poor, But yet I am full of hurtful poison." The image on the right shows the emblem of a lion devouring the sun, which has since been widely reproduced in print and manuscript, usually with the lion depicted in green. The associated epigram explains: "I am the true green and Golden Lion without cares, In me all the secrets of the Philosophers are hidden." Various physical, spiritual, and psychoanalytic interpretations have been put forward for what it symbolizes.

JOH. MICHAELIS FAUSTIJ,

Physici Francfurt. Ordinarij,

Academ. Leopoldin Imperialis Theop

COMPENDIUM ALCHYMIST.

Sive

ANDOR

Explicata & Figuris Jllustra

Das ist / die

delste Gabe

Oder

Ein Güldner Sch

Mit welchem die alten und neuen Philosophi die unvollkommene

Feuers verbessert/und allerhand schädliche und unheylsame Kranckheiten

durch deren Würckung / vertrieben haben.

ieser Edition wird annoch/nebst vielen Kupffern/und über 800. Philosophi

kommenes Lexicon Alchymisticum Novum, und ein vollständiges Register Rerum

Franckfurt und Leipzig /

Verlegts Johan Zieger / 1706.

ABOVE Hand-colored engraving in Johann Michael Faustius's *Compendium alchymist novum* (1706). This image is a variation on the emblem of the *Rebis*, or divine hermaphrodite, reflecting the dual aspects and ultimate harmony of alchemy. The male aspect on the left is represented by fire, the *ouroboros* snake, and tree of the Sun. The female side on the right is represented by water, with the hermaphrodite holding a chalice with birds representing the volatile spirits, and the tree of the Moon at her side.

LEFT Title page from the *Compendium alchymist novum, sive Pandora explicata* (*Compendium of the New Alchemist, or Pandora Unfolded*) by Johann Michael Faustius, Frankfurt (1706). Faustius (1663–1707) was no relation to the legendary Dr. Faust, but was a Frankfurt physician with alchemical interests. The *Compendium* consists of a heavily edited version of *Pandora* (1588) by the physician and alchemist Hieronymus Reusner (although published under the pseudonym of a friar named Epimetheus). *Pandora* was one of the first, and most influential, alchemical printed texts to include a series of woodcut images based on a much earlier manuscript containing emblems full of alchemical symbolism. This edition by Faustius contains hand-colored versions of these alchemical images.

Francis Barrett and "The Magus"

In the so-called Age of Enlightenment, there were deep occult counter-currents flowing across the Western world, and London was one of the centers of learned magical practice. Between 1792 and 1793, while Revolution swept neighboring France, the British capital was the location for the publication of a short-lived periodical entitled *The Conjuror's Magazine, or, Magical and Physiognomical Mirror*, which was dedicated to occult matters. Out of this urban esoteric fog emerged Francis Barrett, described

RIGHT "Heads of Evil Demons, Vessels of Wrath," from *The Magus* (1801), designed by Francis Barrett and engraved by R. Griffith. The three demon heads depicted here are Theutus, Asmodeus, and the Incubus. Theutus is little known, and Barrett likely came across him from a reference in Agrippa's *Three Books* (see page 92), where he is described as the devil who "taught dice and cards."

by one newspaper as merely a "miniature-painter, and an amateur of chemistry," but who advertised his services to instruct initiates in the ancient secrets of magic. Yet, he was better known in his lifetime for his sometimes tragi-comical ballooning adventures reported in the press rather than his magical prowess. We know little about his life, but he certainly left a significant legacy to the development of modern ritual magic in the form of his book, *The Magus: or Celestial Intelligencer* (1801).

The Magus was essentially a compilation of familiar, venerable information rather than an original work of experimental practice. It copied

LEFT Portrait of Francis Barrett, a stipple engraving by English artist Daniel Orme (1766–1837). Barrett is described under the portrait as a "student in Chemistry, Metaphysicks, Natural & Occult Philosophy &c &c."

RIGHT Illustration entitled "Talismans & Magical Images made from the twenty eight Mansion of The Moon," from Francis Barrett's manuscript of *The Magus*, held in the Wellcome Library, London. It is far from an exact copy of the printed version, and was perhaps Barrett's own working grimoire for instructing his pupils before *The Magus* was published in 1801.

heavily from 17th-century English printed books rather than being based on magical manuscripts. Barrett's main sources were Agrippa's *Three Books of Occult Philosophy*, *The Fourth Book of Occult Philosophy*, and the *Heptameron*, a work of magic spuriously attributed to the 13th-century scholar Pietro D'Abano. Yet, Barrett had clearly read much more widely: he was familiar with Solomonic manuscripts circulating at the time, which he had accessed through the London occult bookseller John Denley, whose bookshop could compete with the British Library for its magical riches. Even the famed poet and philosopher Samuel Taylor Coleridge (1772–1834) once borrowed books from Denley for a lecture he gave on "Tales of Witches, Apparitions, etc."

Barrett's book included sections on natural magic, astral talismanic magic, the occult science of "Cabala," alchemy, prophetic dreams, and a discussion on the mysteries of magnetism. He wished his readers "every success imaginable," but warned, "Talk only with those worthy of thy communication—do not give pearls to swine; be friendly to all, but not familiar with all; for many are, as the Scriptures mention—wolves in sheep clothing." Despite all his efforts, and the inclusion of expensive color plates, *The Magus* was a big flop at the time, with one withering review stating, "It would be loss of time to criticize with gravity so weak and ignorant a book." However, decades later, it would come to be appreciated by magicians as a useful general guide.

RIGHT A copper-plate engraving from Robert Fludd's *Integrum morborum mysterium* (*The Whole Mystery of Diseases*), London (1631). Fludd (1574–1637) was a leading English Paracelsian physician and occultist with a degree from Oxford University. His interests were wide-ranging, and at the cutting edge of science at the time. As well as his astrological, alchemical, Kabbalistic, and geomantic interests, he was also fascinated by blood and how it was affected. As shown in this engraving, the four winds were central to Fludd's thinking on the heavenly influences upon human health and fortune. The divine emanations from the Sun imbued the air with the spirit of life, and via angelic influence this was dispersed through the winds that reached the Earth.

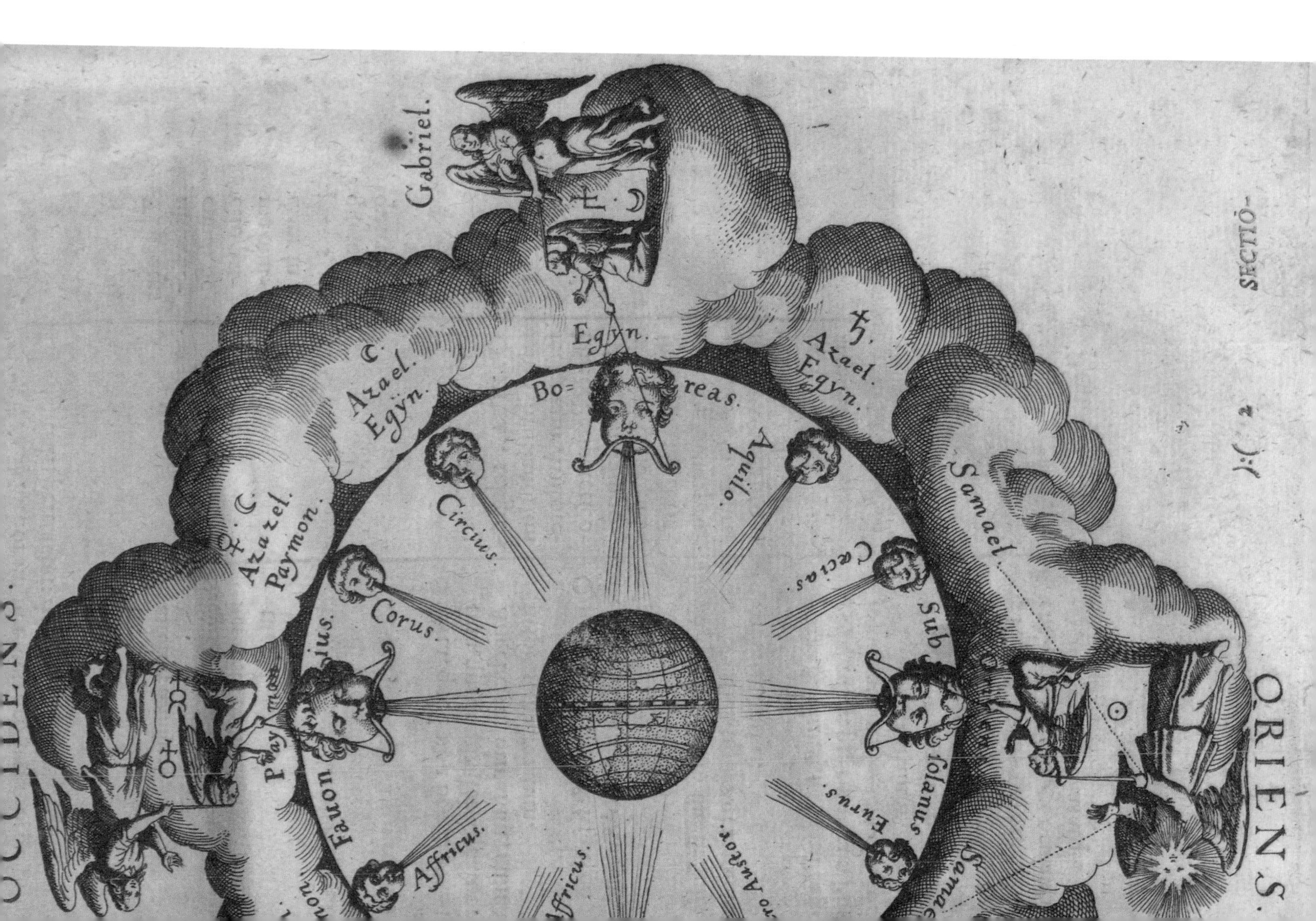
OCCIDENS.
ORIENS.
Gabriel.
Egyn.
Azael.
Egyn.
Azael.
Egyn.
Azazel.
Paymon.
Boreas.
Aquilo.
Circius.
Corus.
Cecias.
Affricus.
Eurus.
Subsolanus
Samael.
Fauon
SECTIO-

GRIMOIRE
DU
PAPE HONORIUS,
AVEC UN RECUEIL
DES PLUS RARES SECRETS.

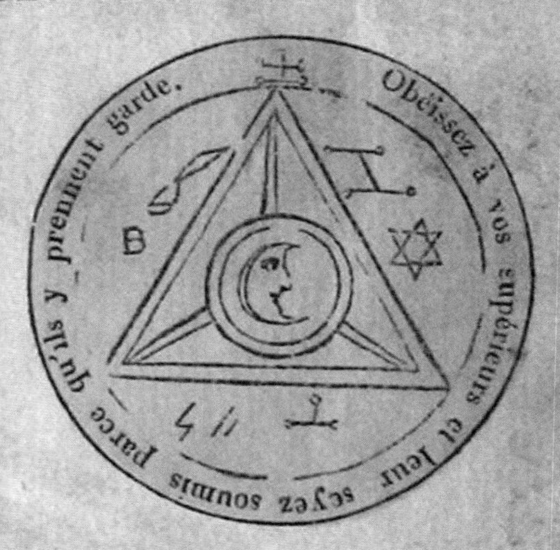

A ROME (1760).

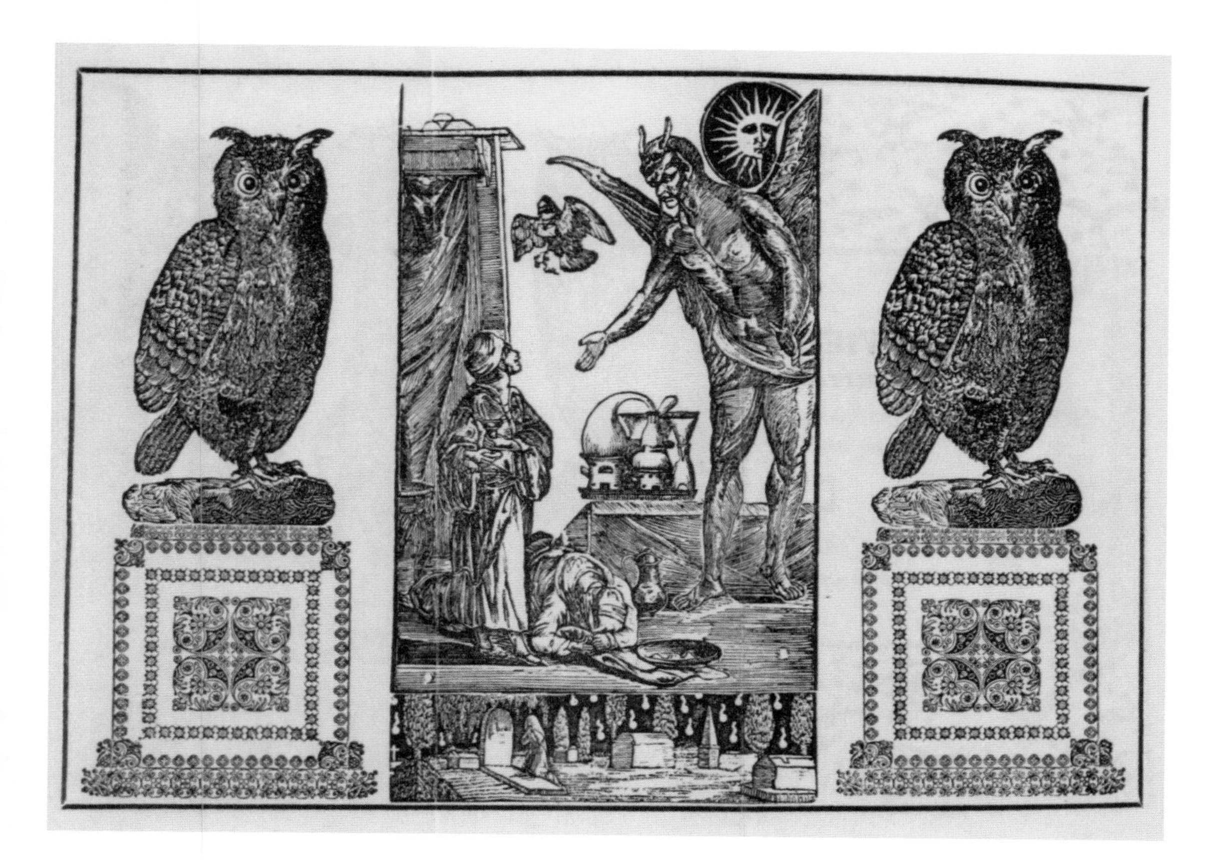

LEFT Title page from *Grimoire du Pape Honorius* (early 19th century). Although the publication details state "Rome 1760," this is one of numerous editions of this cheap conjuration book published in France during the late 18th and 19th centuries. While the medieval *Sworn Book of Honorius* (see page 57) was attributed to a fabled 4th-century Egyptian magician, this grimoire was spuriously stated to be the work of Pope Honorius III (d. 1227). Manuscripts with the same name had been circulating in early 18th-century Paris, and it is likely this popular print grimoire derived from one such text. Its content is couched in pious terms, but it includes the conjuration of demons and Lucifer himself (but only on a Monday).

ABOVE *Trésor du Vieillard des pyramides, véritable science des talismans avec la Chouette noire* (*Treasure of the Old Man of the Pyramids, True Science of Talismans with the Black Owl*), Lille (1830). This cheap French grimoire contains a series of images of sigils, or talismans, to be created in metal or inscribed on virgin parchment with magical inks. The book relates that they were obtained from an old Arab man by a French soldier on expedition in Egypt. The man hid the soldier in a pyramid and showed him the secrets of magic from ancient texts. While in the pyramid, the old man also gave him a manuscript in multi-colored hieroglyphs that instructed on the ritual for obtaining the eggs of the marvellous Black Owl, and how to hatch them. The possession of such an owl was said to unlock ancient powers.

يرالسلام نوح
عليه السلام عليه
السلام

CHAPTER FOUR

Manuscript Culture Thrives

The world of early print magic was largely one in black and white. Hand coloring of printed woodcut and plate images did happen commercially, such as for herbals, and was often done by women and children, but it was an expensive production process for the luxury market, and such books were only produced in limited numbers. Few magical and alchemical publications received this treatment. So, the handwritten manuscript form continued to be a vital medium for the expression of the esoteric in color. This was the case particularly with alchemical works, where different colors and the use of gold leaf, for instance, had symbolic as well as artistic qualities. The very act of writing, transcribing, and illustrating was considered a transformative process and a performative art that embodied the magician in the physical manuscript, creating a bond of occult sympathy that print could not replicate.

Of course, there were other reasons for the continued enthusiasm for manuscript magic. Most printers, even if they had the freedom to do so, were reluctant to print some aspects of magic, such as demonic conjurations, or spells for harmful or sexual purposes. Then there was the venerable desire to control and restrict magical knowledge. While one wing of occult intellectualism was all for the promotion of the hidden workings of the world, the other was keen to maintain a tradition of secrecy and fraternity regarding alchemical or magical wisdom and experimentation.

In an age of expanding popular literacy, particularly in Protestant states, new manuscript grimoire traditions took off in Germany and Scandinavia that thrived well into the 19th century. Most of the surviving examples are simple black-ink books compiled and copied, sometimes falteringly, by cunning-folk and aspiring magicians. A few were far more ambitious, however, containing folk art and calligraphy of the highest quality.

In some regions, the continued flourishing of manuscript culture was more simply a matter of very restricted access to printing. There was only one printing press in Iceland until the mid-18th century, for example, and it was owned by the Church, which obviously had no interest in promoting occult matters. Yet, with the increasing availability of cheap paper on the island, a vibrant manuscript tradition flourished, leading to the circulation of ancient sagas, romances, tales, poetry, and magic.

LEFT Title page from *The Tree of Knowledge* (*Ets ha-Da'at*), a parchment codex of magic in Hebrew (16th century). This little book contains 125 magic spells for protection, healing, and acquiring knowledge. Its "discovery narrative" tells how a man named Elisha obtained the contents from an old book he found in Venice and from his visit to the Holy Land. It shows the continued vibrancy of Jewish manuscript magic in Italy during the early modern period.

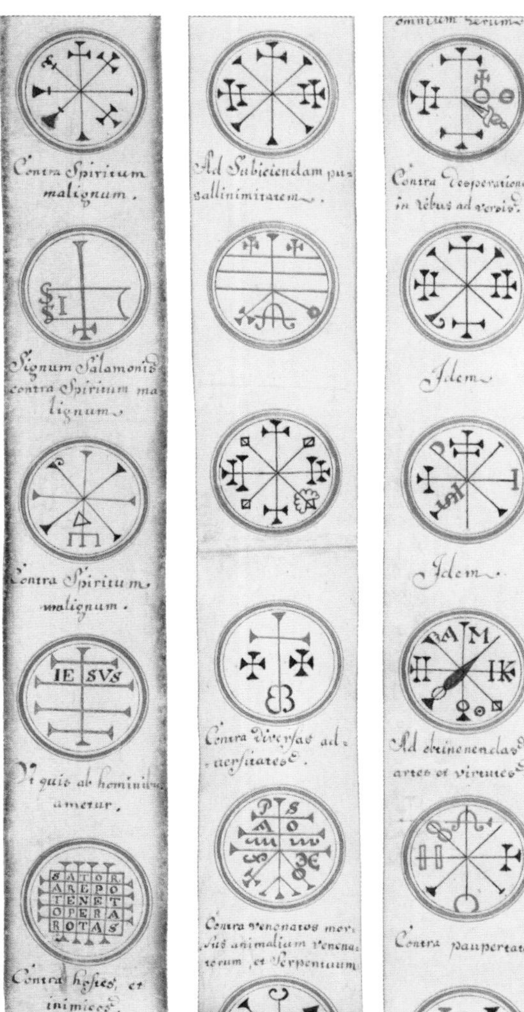

RIGHT Three sections of an English amulet roll containing magical seals, on vellum (17th century). It was unusual by this time for rolls still to be created in Europe, rather than using the book format. On one side are Latin prayers, and on the other (shown here) 63 magical seals against such dangers as snakebites, wounds, evil spirits, and even poverty. This roll was used either as a protective amulet, a magical manual, or both.

In German-speaking territories, the 18th-century Faustian *Höllenzwang* publications we heard about in the last chapter (see page 97) were almost certainly an opportunistic response to the popularity of such Faust texts circulating in manuscript form. A collection of German magic manuscripts auctioned around 1710 already included eight Faustian *Höllenzwang*. An analysis of their contents, and others, shows that they were partially derivative of earlier Solomonic ritual works, but also were sometimes prefaced by a first-person address from their pseudo-author—Dr. Faust. In one example, Faust states that he had read many occult books from an early age, including a book full of conjurations for summoning spirits:

> "I tried them only for an experiment. Nevertheless, I became aware that a mighty spirit, named Astaroth, presented himself before me, and asked me wherefore I had cited him. Then, hurried as I was, I did not know how to make up my mind otherwise than to demand that he should be serviceable to me in various wishes and desires."

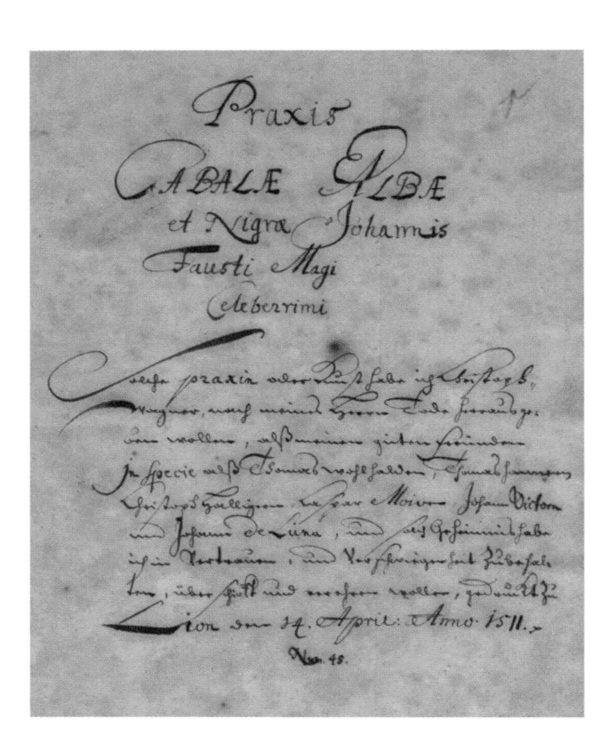

Praxis
Cabalæ Albæ
et Nigræ Johannis
Fausti Magi
Celeberrimi

Lion den 24. Aprill: Anno 1511.

LEFT Title page of *Praxis Cabalae albae et nigrae Johannis Fausti Magi celeberrimi* (*The practice of the white and black Cabala of John Faustus the famous Magi*), a paper manuscript in German script (18th century). One of numerous variations on the Faust *Höllenzwang* genre of grimoire (see pages 151–155).

One of the reasons these books were so desirable was the lively popular interest in treasure hunting during the 18th century. Court records describe groups of men obtaining a Faustian manuscript in order to call up and control the spirits and demons that were thought to guard buried precious stones, coins, and metals.

Faust grimoires also heavily influenced the Scandinavian tradition of manuscript *svarteboker*, or "black books," which circulated in their hundreds across the region during the 18th and 19th centuries. They were owned by budding magicians from most levels of society in Denmark, Norway, and Sweden, but in the 18th century the authorities were particularly concerned about soldiers, who possessed these black books to make pacts with the Devil in order to protect themselves from harm. While some *svarteboker* did contain instructions for making a pact, it is not a dominant theme in the dozens of surviving examples in archives today. A number of them provide conjurations for obtaining buried treasure, but otherwise they contain a mixture of simple spells, charms, and remedies for everyday health and protection.

GLOBAL TRADITIONS

Away from the European magical milieu, popular manuscript magic also flourished across parts of Africa and Asia from the 17th century onward. Although deeply influenced by Christianity, Islam, Hinduism, and Buddhism, these manuscripts also show a fertile mix of regional cultural and indigenous religious influences. This was a period of expanding European colonization in the Middle East, Africa, and Asia, and the many editions of overseas magical texts in European collections today often derive from conquest one way or another. The literary magical heritage of numerous cultures was looted, confiscated, purchased as colonial curiosities, sold as commercial objects, or obtained through missionary zeal to understand or suppress indigenous religions. At the time, Asian and African magical cultures were often, although not always, treated as "primitive" examples of an early stage of humanity. There was a glaring double standard, of course, for at the same time similar magical manuscripts were created, traded, and *used* in educated European occult circles, as well as in popular culture.

Our knowledge of magical manuscript culture in west African history is limited. One of the largest known collections of written talismans and treatises with instructions on how to employ them, actually resides in the Danish Royal Library. In origin, this collection was possibly appropriated by an army of the Ashanti kingdom that attacked the predominantly Islamic Dagomba tribe in northern Ghana in 1744–1745. Several decades later, it came into the hands of officials in the Danish colonial settlement of Christiansborg in southern Ghana. Sometime in the early 19th century, the collection was transferred to Denmark. Islam was not a significant religion among the Ashanti, but Islamic forms of written protection based mostly on the Qur'an came to be highly prized. The Danish collection shows three main types of written magic: the application of passages from the Qur'an, including erasure magic; talismans referring to the life of the Prophet; and occult Arabic word squares and numerology. They were used for military purposes, protection from the evil eye, to kill enemies, promote good fortune, and heal sickness.

The manuscript copying of medieval Arabic–Islamic magic books remained prevalent in eastern Africa. Copies of the *Kitāb al-Mandal al-sulaymānī*, a late medieval grimoire that provided conjurations to call up named *jinn* in order to exorcise people, from the 18th and 19th centuries have been found in the archives of Ethiopia and Yemen. One early 20th-century example was written in a simple, lined notebook manufactured in Britain. There was a particular colonial interest in Christian Ethiopian manuscripts with thousands plundered and traded during the 19th and early 20th centuries, which are still held in public and private collections today. Among them are numerous works of magico-religious significance. Collectors were particularly fascinated with the distinctive amuletic scroll texts written in *Ge'ez*, an ancient "dead" language that, like Coptic in Egypt, remained the official liturgical language of various Ethiopian Christian churches. They were

RIGHT Section from an Ethiopian healing scroll, written in *Ge'ez* script on parchment (19th century). Prayers can be seen written in black ink, while section headings, the names of the Holy Trinity, and the client's name are in red ink. The talismanic image shown here is based on an eight-pointed star associated with Solomon's ring or seal in Ethiopian tradition, which is a common motif in such scrolls. The face is possibly that of a protective angel.

ABOVE Icelandic magical manuscript on paper (1860). A copy produced from older manuscripts by a scholar named Geir Vigfússon (1813–1880), of Akureyri, who also copied other non-magical manuscripts. While the front page refers to the Hebrew alphabet, and includes pseudo-Hebraic characters, the contents also use both runic alphabets and distinctive *stafir* (sigils or characters), including a Seal of Solomon and a Seal of David.

mostly written on parchment by the *debtera*, who were (and are) a cadre of Orthodox holy men making money from healing and offering magical services, as well as providing religious duties. The small scrolls they sold were kept in cylindrical leather containers, or were tied up with cloth and worn on the person for protection against the evil eye and ill fortune, while larger scrolls consisting of parchment leaves stitched together were sometimes hung on walls unfolded.

Moving across to Asia, manuscript handbooks of *yantras*—geometric talismans based on the principles of ancient Hindu astrological science and tantric symbolism—were in widespread circulation not only in India but also across South Asia. Often associated with specific deities, *yantras* were written on bark, reproduced on paper, or cast in metal, and were used not only for protection but also for exerting influence over external forces, whether human or otherwise. Some were produced in conjunction with written Sanskrit *mantras* or sacred utterances. They were inserted into statues of gods, concealed or hung in the home, buried, kept in a locket or amulet container, worn on the person, or sometimes even ingested. A tradition of tattooing *yantras* also developed as an aspect of religious devotion and occult protection across Southeast Asia, particularly in Buddhist cultures. As with early-modern Solomonic texts, there is a considerable emphasis on the importance of triangles, hexagrams, pentagrams, and circles. While examples of *yantras* from early antiquity can be found, the popularity of their application in more recent times, as well as the interplay between manuscript copying, cheap print engraving, folk art interpretation, and cross-cultural influences, means that over the last couple of centuries the corpus of *yantra* designs has become a more free-form aspect of art magic—more *yantra*-like than representing adherence to ancient purity. They have also proved readily adaptable to aspects of modernity. *Yantras* representing the Hindu monkey god *Hanuman*, for instance, are frequently sold today to hang in motor cars to protect against accidents.

The corpus of Lontar palm-leaf manuscripts of Bali, mostly dating to the 19th and early 20th centuries, contain magic texts along with religious, medical, genealogical, and cultural matters. Sanskrit Indian religious influences are strongly evident, and the writings also show Malay, Indonesian, Javanese, and colonial Dutch linguistic and cultural elements. Lontar manuscripts do not survive for more than a hundred years or so in the natural environment of the region, and so in the past there must have been a systematic tradition of meticulous copying as libraries slowly decayed. Most of the surviving examples have thankfully now been digitized by the government. To turn Lontar leaves into a writing surface, the leaves were trimmed, dried for several days, and then given a lengthy soaking and/or boiling in a herbal mix to give them strength. Then they were dried very slowly and pressed for days to flatten. The edges were treated with a dye to protect from insects. A sharp iron tool was used to engrave the leaves with writing, and the text was then highlighted by rubbing in lamp black (a finely powdered black soot). The resulting long, rectangular leaves were held together with a piece of cord to create Palm manuscripts.

A survey of over 1500 such Balinese manuscripts revealed that 30 percent contained *mantras* for healing or were solely concerned with magical matters, and most were obtained from magicians known as *balians* or from holy men (*pandandas*). As well as the usual healing and protective magic, there was also a dark tradition of causing harm by calling upon the

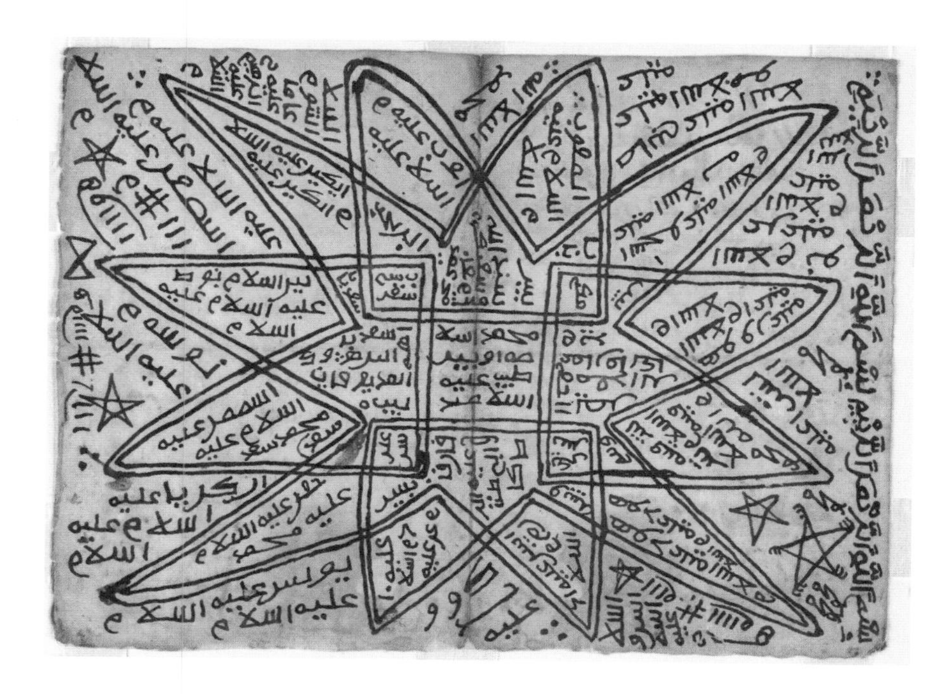

LEFT Sudanese Islamic amulet, in ink on paper (19th century). This was an amulet to be worn against the power of the evil eye. It is from a miscellaneous collection of Arabic esoteric manuscripts (one of numerous in European libraries), including an older Arabic alchemical text, and another Sudanese charm consisting of letters from the word "Mohammed."

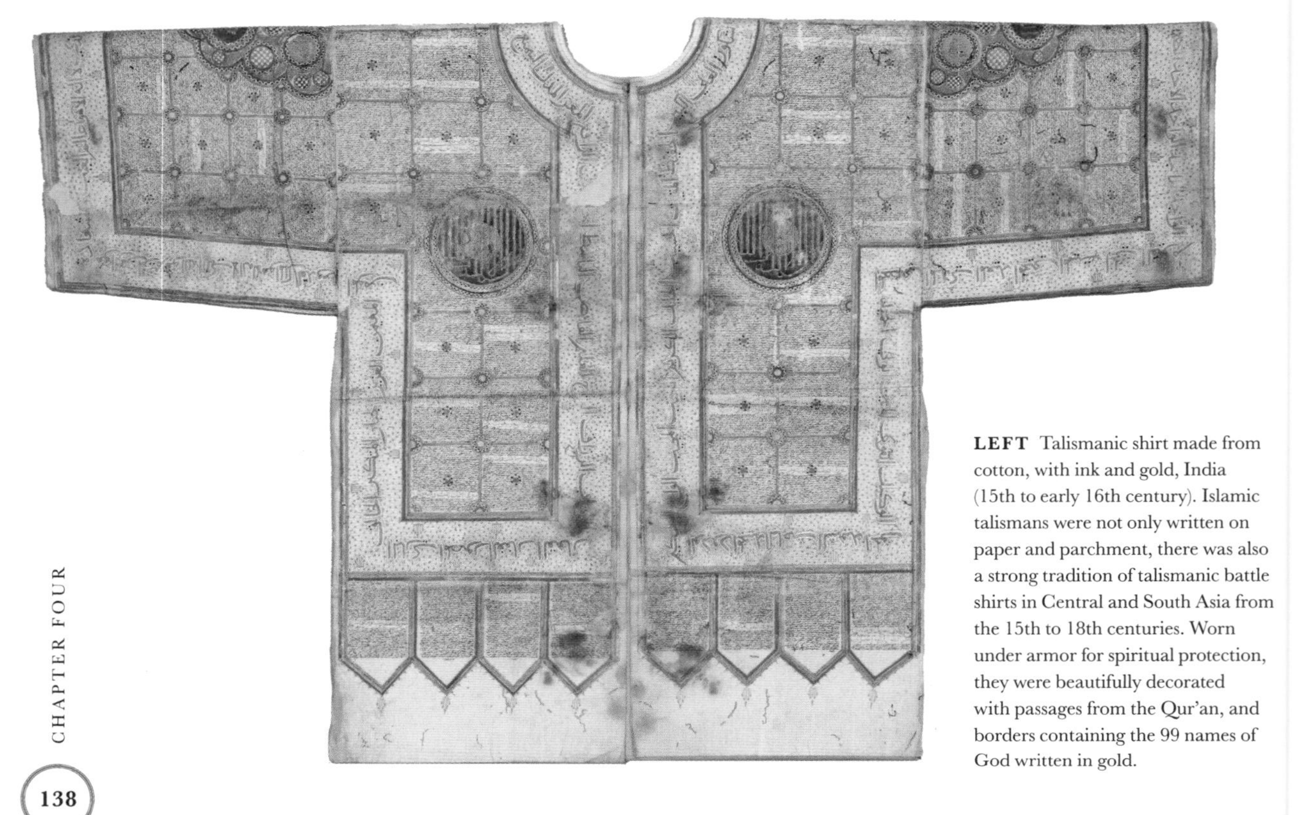

LEFT Talismanic shirt made from cotton, with ink and gold, India (15th to early 16th century). Islamic talismans were not only written on paper and parchment, there was also a strong tradition of talismanic battle shirts in Central and South Asia from the 15th to 18th centuries. Worn under armor for spiritual protection, they were beautifully decorated with passages from the Qur'an, and borders containing the 99 names of God written in gold.

Indonesian deity Durga, the goddess of black magic and demons—a far cry from the original, benign mother goddess Durga of ancient Hinduism. One form of harmful magic, known as *papasangan*, required the operator to summon demons and activate their evil powers by writing accompanying incantations. The resulting manuscript was then buried or concealed in and around the house of the intended victim. There were *balians* who also offered their services to find and dispel these curses.

In northern Sumatra, now part of Indonesia, several closely related ethnic groups, known collectively as the Batak, developed another writing tradition using tree bark on which they wrote their own script in a black ink made from charcoal. These *pustaha*, or tree-bark books, sometimes with ornate, carved wooden covers, were collected as colonial artistic curiosities by Europeans from the 18th century onward and taken back to Europe. The large collection held by Leiden University was donated by Herman Neubronner van der Tuuk (1824–1894), a linguist who was employed by the Netherlands Bible Society in the 1850s to journey to Sumatra to translate the Bible into the Batak Toba language. Van der Tuuk's strategy for learning Toba involved collecting and attempting to translate *pustaha* obtained from local *datu*, or magician-priests. Although not deeply religious, van der Tuuk was disturbed by some of the content he had to read in this endeavor, and one time declined to look at a *pustaha* full of sorcery owned by the raja or prince of a Batak village, stating that "to consider its purchase was unthinkable."

The *pustaha* corpus mostly consist of a blend of magic, mythic stories of gods and monsters, astrological tables, divination, and medicine. There were instructions for reading omens from the shape of clouds, or divination by the entrails of a chicken, accompanied by an illustration of a chicken in black and red ink. As for magic, one *pustaha* author advised *datu* pupils to study its contents very carefully as "however strong may be a fence made of hardwood, still stronger is a fence of magic." There was an incantation in one book to remove harm from a foetus in the womb, for example, while a more powerful spell concerned the activation of a guardian warrior spirit, known as a *pangulubalang*, which was represented as a carved figurine, so that it would protect the community from human or spiritual attackers.

Magic and divination in neighboring Malaysia, demonstrate the complex range of religious and cultural interactions in Southeast Asia, but the old Arabic traditions of magic described in previous chapters (see pages 54–56) certainly form the basis of the 18th- and 19th-century collections of Malay magic books. Islam was introduced to the Malay peninsula by Arabic traders back in the early medieval period and by the 16th century, Sunni Islam had become the majority religion. The Malay manuscripts, mostly written on European-made paper, but also *khoi* paper made in Thailand from the bark of a native tree (*Streblus asper*) since the 17th century, are rich in imagery, some of it pre-Islamic in nature. Depictions of Solomon's seal as a pentagram are a significant presence, and the pentagram is known as *tapak Sulaiman* (Solomon's hand or footprint) in Malay. Various manuscripts instruct on the use of the *selusuh Fatimah*, or Fatima's talisman, for childbirth, which consisted of a three-by-three grid containing Arabic letters drawn on a piece of paper, and attached to the right arm of the pregnant woman with white thread. According to the texts, it was said to have been given to Muhammad's daughter Fatima by the angel Jibreel (Gabriel). Yet, Malay manuscripts also reflect the fact that magicians borrowed from other ethno-religious traditions, particularly Thai magic to the north. Several 19th-century examples include talismanic designs of Thai origin, including one that has a drawing of a human form with a Thai talisman on its chest and Arabic script surrounding the image. It has been suggested that Hindu elements in Malay magic texts might have derived more recently from Java rather than being a direct survival from pre-Islamic times.

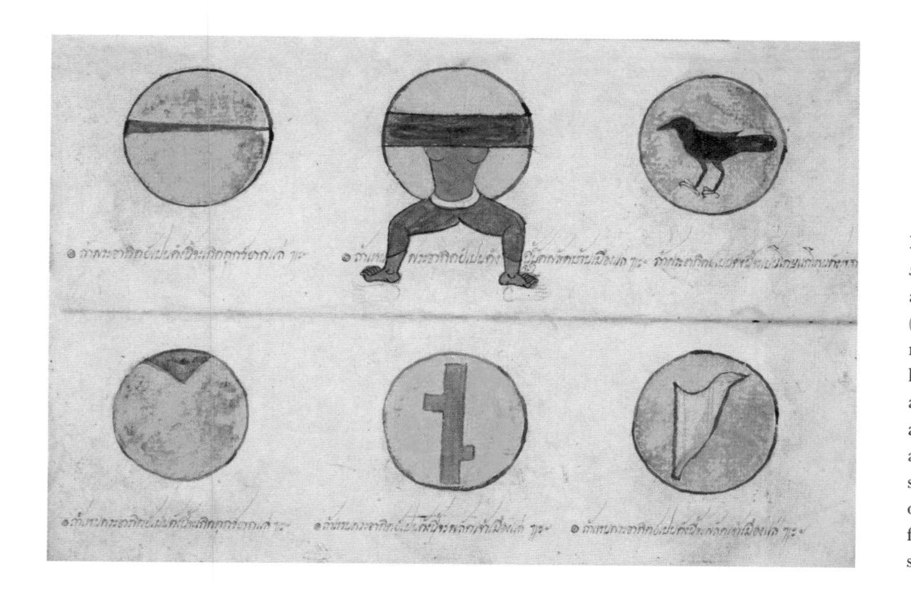

LEFT Section from the *Tamra phichai songkhram* (*Manual of Victorious Warfare*), a paper folding book, Thailand (1800–1880). First compiled as a guide to military strategy in the late 15th century, later versions focus on broader divinatory aids, including astrological observations and the interpretation of cloud forms. They also contain *mantras* and rituals for military success. This page concerns the appearance of the Sun. The text relating to the human figure reads: "When thus the Sun appears, separated from his native city each will be."

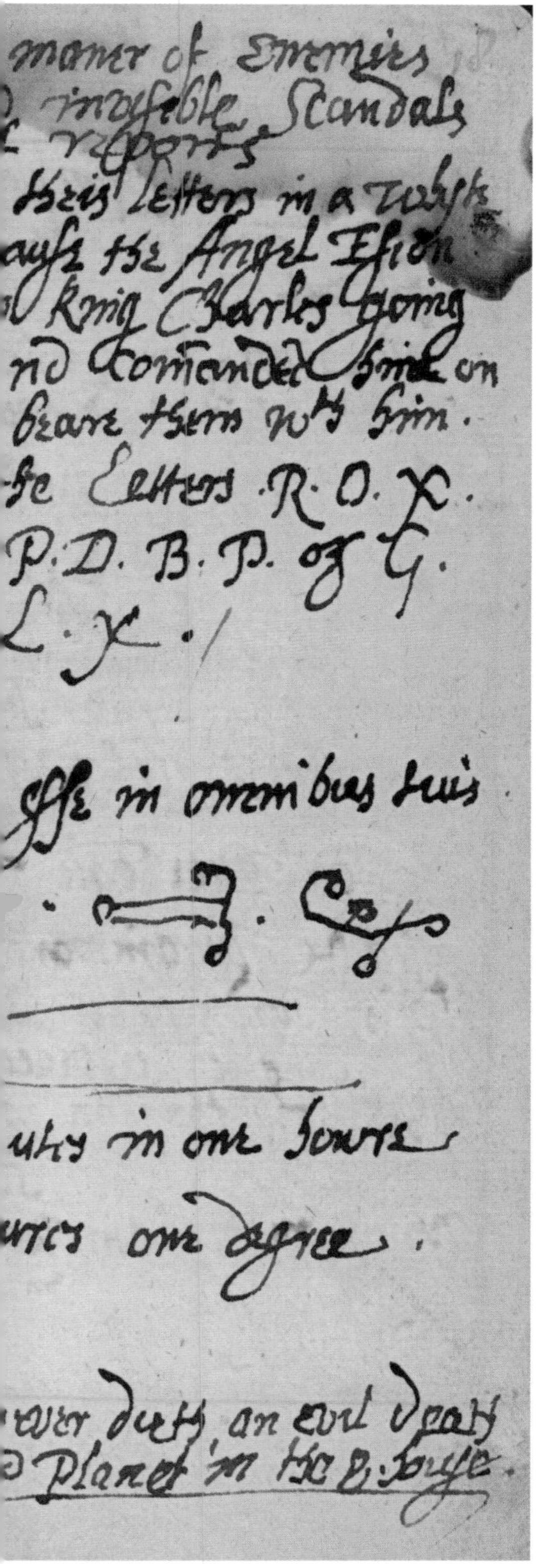

LEFT Book of miscellaneous magical charms, on parchment, English (early 17th century). The author has been identified as the lawyer and bibliophile Robert Ashley (1565–1641). He compiled this work mostly from other manuscript sources, but also from a few printed books, including Reginald Scot's *Discoverie of Witchcraft* (see page 93). The page on the left begins: "To Speak with Spiritts Call their names Orimoth, Belmoth Lymocke and Say thus. I coniure you by the names of the Angels + Sator and Azamor that yee intend to me in this Aore, and send unto me a Spirite called Sagrigit that doe fullfill my comanding and desire and that can also understand my words." The page on the right begins: "Against all maner of Enemies visible and invisible, Scandals and evil reports. Beare with thee theis letters in a whyte linen Cloth: because the Angel Esion did beare it to King Charles going to Battayls."

make an eye on the wall or in pchement, & strike him one
then take an naile of Latton of the waite of a penye x
of ewe the hed & the steale of box, yt is a malett, & then m
eye ij circles & write in the firste of the ij Jesus Saluator,
lesser circle write, Jesus Sciens reru occultaru & manifestaru
purgator, omnis nōie prioso, then say this charme, I coniure a
lokers on this eye, & all them yt in this thinge be guiltye of
belonginge of this eye, I coniure them by the vertue of the
of the sonn & of the holy ghoste, & by all the names of go
& crit + & by all the appostles, & by all the evangelists &
& confessors & by the holy eliments, & by + mary the mo
our lord Jesus Christ, and by all their workes, yt all them
guiltye of suche thinge wch is gone, & all they that belonge
so faste, strike it on his eye, by the vertue of the holy names
lord Jesus Christe, before said, yt it never sere till his eye
or give answere & for them be brought againe, & god for his
mighte, & right as I smite this naile one this place yt we n
yt all vertue yt is in these wordes aforesaide, may torne to
to confusion, & while thou smitest saye Xabat, & Xabas, sei
Reatonay sehare Reatony facite apprere qui illam Rem, furatus
qua querimus, & thou shalt se the righte eye water & if wat
smite the eye againe wth the hamer & begin this so ofte ti
last fidinge of the thefte & begin this charme on the firste
the mone or in the laste qter of the mone when thou wilte,
followes the forme of the eye as it ought to be made;

a b r a c a d a b r a
a b r a c a d a b r
a b r a c a d a b
a b r a c a d a
a b r a c a d
a b r a c a
a b r a c
a b r a
a b r
a b
a

sicut
deleo hanc
literā de isto
scītissimo nōie de
abracadabra ita p
virtute huius sacratissimi
nōis doleat morbus et dolentium
a IB, In nōie pris et filij et sprit
Pr deleat te morbū deus filius + deleat te
morbū Amen
orōn sprit scts + deleat te morbū Amen

+

For the tothacke Ile · Corbe · hor · horss · gaubell · X · peboxtdm

This abovesaid, must be written in a little pece of pch
and at the scrapinge out of euery line, you muste say ot
I thinke yt were mought, yf only the eleuene letters, abra
bra, wer written & scraped out, singulatim, you must
begin at the furthest or lowest line, a, and so ascende vp

dnutt txtebotr
b dn ttr tyt
ot xo vxtm
dbsnitou vtreh
mxthro xo bnd

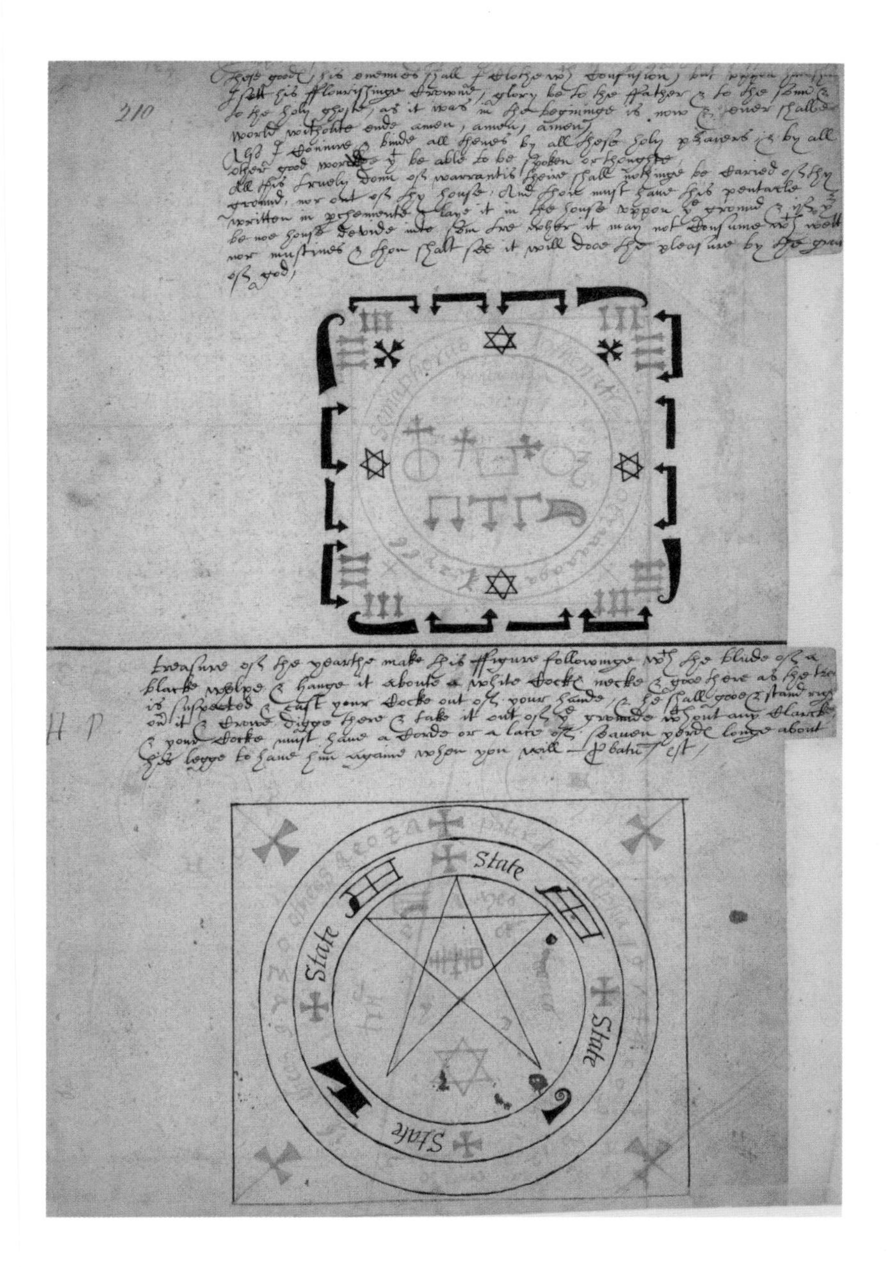

LEFT AND ABOVE Solomonic compilation of magic, with instructions for invoking spirits, written in English, Latin, and some Hebrew (c. 1577–1583). The image (left) concerns an "experiment" called the Eye of Abraham, used to make thieves confess their guilt. As part of the ritual the magician is instructed to draw the picture of the eye on a wall, or on a piece of parchment, which is then hung on a wall. A charm is recited that begins: "I Coniure all the lokers on this eye, & all them that in this thinge be guiltye of & is behouldinge of this eye." The top image (above) concerns a long adjuration to protect the home from thieves. It was to be reproduced on parchment and placed on the ground in the house concerned. The bottom figure was used to find treasure buried in the earth. It was to be drawn on parchment with the blood of a black whelp, and then placed around the neck of a white cock.

ABOVE Alchemical manuscript on paper, German (early 17th century). One of several colored drawings from an alchemical poem. The image is more obscure than most in its meaning, but certainly vibrant in its depiction. It shows an alchemist waist deep in the ground, an interlaced rope around him, and his books and apparatus laid before him. A boy's head appears above the ground. Other more easily interpreted images from the manuscript include a green dragon being slayed, the four elements represented in animal form, and an alchemist in his laboratory.

RIGHT Illustration and epigram from Elias Ashmole's edition of the *Theatrum Chemicum Britannicum* (vol. 2), manuscript on paper (17th century). In 1652, Ashmole (1617–1692), a well-connected antiquarian and one-time alchemist, published a compilation of alchemical discourses, verses, and emblems, taken from an array of rare manuscript sources dating back several centuries. This image with epigram is one of several interleaved sketches and hand-colored engravings. The document illustrates the interplay between print and manuscript at the time, when occultists added their own manuscript notes, addenda, and illustrations to printed books, just as they had to manuscripts.

Astra regunt homines
Quod est superius est sicut inferius
Fraxinus in Silvis pulcherrima,
Talpa in Terris operosissima,
These Hieroglyphicks vaile the Vigorous Beames
Of an vnbounded Soule: The Scrowle & Scheme's
The full Interpreter: But how's conceald.
Who through Ænigmaes lookes, is so Reveal'd.
T. Cross sculp:
T:W:M:D:

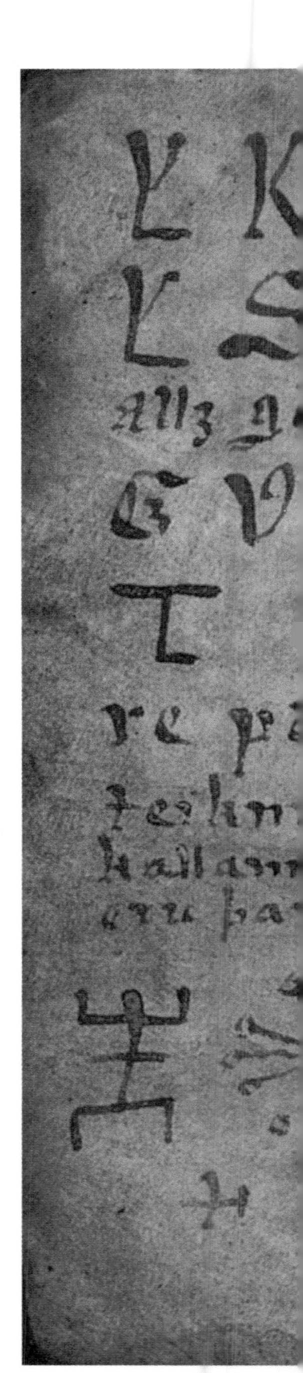

ABOVE AND RIGHT Cover and binding of an Icelandic manual of magic, written in Icelandic and Latin on animal skin (c. 1670). Known as the *Galdrakver* (*Book of Magic*), this is one of the earliest surviving Icelandic grimoires, and was written on skin at a time when paper was becoming widespread on the island. Its contents, like other Icelandic magic books, were concerned with spells for protection, healing, and counter magic rather than spirit conjurations, and included the usual Greek, Latin, and Hebraic letters and characters found elsewhere in European magic books, along with distinctive runic symbols. Yet, the mainland Scandinavian Faust and Cyprianus traditions are little evident. Such manuscripts featured prominently in 17th-century Icelandic witch trials that mostly targeted male magicians and cunning-folk. A few of those found guilty of possessing magical manuscripts had the pages burned under their noses while being flogged.

Ægis hialmur hn̄ ꝥ
gioraſt a blÿ ꝫ þrickia
ꝫ enni sey þarm̄ a uoni a
oþin synnū ad hn̄ mæti
hn̄ ꝫ mæntu han̄ hr/uiña
(hn̄ ar so f hn̄ epty si/97)

11

tora
ef stauū
ilgia þeſsu
sarletri og
lta framma

Yōkai

There is a rich Japanese art history of depicting demons, known as *yōkai*. Old *yōkai* encyclopedias, which described their powers and characteristics, were not unlike the lists of demons or *jinn* in the European and Arabic magic tradition, but there was a particular emphasis on the artistic representation of each *yōkai*. The most influential of these encyclopedias was *The Illustrated Night Parade of One Hundred Demons* (1776) by the Edo-era artist Toriyama Sekien (1712–1788). It was one of four volumes about the *yōkai* in a series known as the *Gazu Hyakki Yagyō*—a Japanese folkloric term for the wild nocturnal processions of demons and ghosts, sometimes led by a chief *yōkai* known as *Nurarihyon*, which were thought to occur in villages and towns across the country. On such nights it was best not to stray outside for fear of capture or even death. Among the horde, there was the *akanamé*, or "filth licker," a creature that licked the scum out of dirty bathtubs, for example, and the *yamauba*, or "mountain hag," who brought the snows in winter and blossom in spring. Then there was the *oni*, an ogre-like demon bearing horns, who had a strong dislike for humans. His devil-like image can be found in "Hell Scrolls" from the 12th century that depict the various Buddhist realms of Hell, or *Narakas*, such as "the inferno of excrement."

Sekien drew upon a long history of *Hyakki Yagyō* depictions on paper scrolls, including

BELOW An illustration from Toriyama Sekien's encyclopedia *The Illustrated Night Parade of One Hundred Demons* (1805 edition). This depicts a very hairy, old female *yōkai*, known as the *ouni*, with a demonic grin. Sekien included no explanation with the image and the *ouni* is not a recognized category of *yōkai*. However, it is very similar to the hairy hag-like *wau-wau* found in earlier Japanese demonologies.

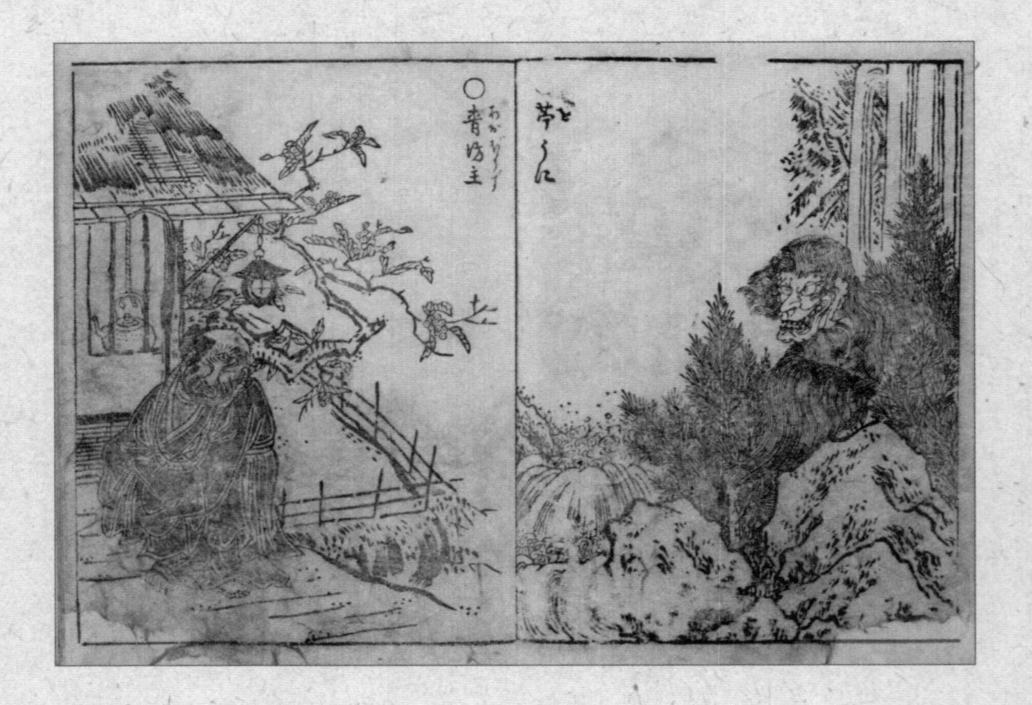

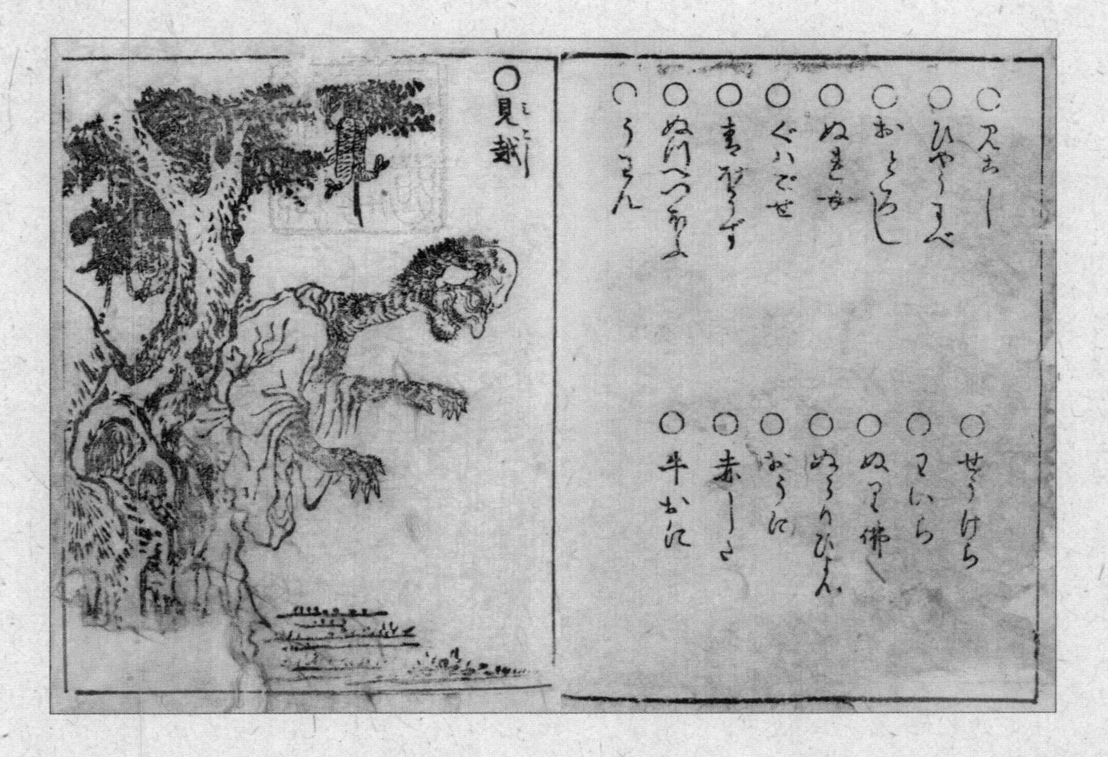

ABOVE An illustration of the *miage-nyūdō* from Sekien's *The Illustrated Night Parade of One Hundred Demons* (1805 edition). The *miage-nyūdō* is a type of *mikoshi-nyūdō*—a bald-headed, monk-like being with an extendable neck. They are usually encountered by lonely travelers in the mountains and forests. At first glance, they appear to be a fellow human but suddenly they grow abnormally tall and threatening.

medieval Chinese images of demon processions. The oldest surviving example is the Shinju-an scroll, dating to the early 16th century and preserved in a Kyoto sub-temple. It depicts a fantastical scene of *tsukumogami*—tools and objects that become animated spirits in Japanese folklore. But the works by Sekien and others, in the late 18th and 19th centuries, broke new artistic and cultural ground by using the structure of the book rather than scrolls. There was one page for each demon, but there was no linking narrative at all. Each image told its story, and can be seen as an independent unit of knowledge for subsequent authors and artists to repurpose. Sekien's visual style, as well as the way in which the *yōkai* were cataloged, inspired numerous 20th-century Japanese authors and artists.

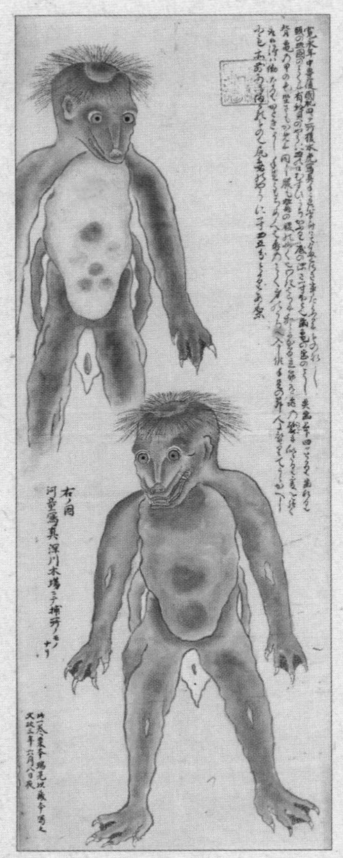

RIGHT Illustration of a *kappa*, in ink on paper (probably 19th century). The *kappa*, or river child, is one of the best-known *yōkai* in Japan, today. They live in ponds, swamps, and rivers, and are generally slimy, scaly, and green, with a carapace on their backs. They are about the size of a child, but much stronger than a man, and are believed to drag horses and children into the watery depths.

ספר חמישה חומשי תורה ס׳ הלבנה א׳ ספר המזלות א׳ ס׳ מפתחות
שלמה עם חכמתו א׳ וספר הניסיונות אחד
ואת המאור הגדול לממשלת היום והלילה

1611

RIGHT Two images of infernal spirits from ***Iohannis Faustii Magia Naturalis Et Innaturalis** (John Faustus's Magic Natural & Unnatural)*. This German manuscript was purportedly published in Passau in 1612, but is almost certainly from the 18th or early 19th century. This is one of the most beautifully illustrated Faust *Höllenzwang* grimoires (see page 133). The top image depicts one of the Devil's lieutenants, Mephistopheles, in human form, and the demonic bull below is the appearance of the infernal spirit Aciels, when conjured.

LEFT Diagram from the *Sefer Maftea Shelomoh* (*Book of the Key of Solomon*), a Jewish magic book written on paper in Hebrew, with passages in Italian, Latin, Arabic, and ancient Greek, Amsterdam (17th to 18th century). It contains a series of instructions, images, sigils, seals, and characters for conjuring and binding spirits, and rituals for other diverse purposes, including escaping from prison.

10. Nun folget das Sigel, welches zu Schatzgraben erfordert wird.

S. Trinitas. ✠ Alpha et Omega, Tetragramaton, Emanuel, Jesus, Christus, Sotir, Athleter, Adonay, Heli, Sabaoth, Messias, Abrakola, Abrakol, Abraka, Abra, Aber ... Agla, Lux, Verbum, Flona, Lux Imago, Sponsus, Eaca,

✠ Hagios, o theos is Kirios Athanatas Imas, Eleison, Panthera, Aimulo, molon, onaytheon in Excelsis ons taleron Flamandum Is rohas Hya ... Almiseron Or eysleon Anayon Loy Ory & Lamathon Amen.

✠ Jesus autem transiens per medium illorum ibat

Jesus Crucifixus
Alpha et Omega
Sabaoth Messias

Gloria Lux

Chaos
Agla

Eloy

Agla

Eloy
Elohim

Lomathum

Adonay

Vinces

LEFT Page from *Praxis Cabalae albae et nigrae Johannis Fausti Magi celeberrimi* (*The practice of the white and black Cabala of John Faustus the famous Magi*), a paper manuscript in German. It is stated as written in 1511, but is dated to the 18th century. This is another Faustian manual used for conjuring diabolic spirits to attain knowledge, wealth, and health. The circle contains the usual elements one might find in grimoires from the late medieval period, including planetary signs, pseudo-script, and magical Hebraic divine names such as "Agla" and "Tetragrammaton."

ABOVE Drawing from *Praxis Cabalae albae et nigrae Johannis Fausti Magi celeberrimi* (*The practice of the white and black Cabala of John Faustus the famous Magi*). This chapter concerns Mephistopheles, who was sometimes represented as a demonic being, or sometimes in the guise of a Franciscan friar, in the story of Faust's pact. This image may have been inspired by a well-known copper-plate engraving by the Dutch printmaker Christoffel van Sichem I (c.1546–1624), published in 1608, which depicts an encounter between Faust in his finery and Mephistopheles as a tonsured cleric. The following page in this manuscript includes two sigils or characters of Mephistopheles to be used when making a pact.

ABOVE The *Dochder Johannes Faust III Fager Hollen Zwanck Schwartze Magie und Kunstz und Wunder Buch*, German manuscript book with black ink throughout (18th century). Purportedly written in Lyon, France, this 24-page book is written in German and in code. Part of it is supposedly from the *Sixth Book of Moses* (see page 174). As well as the enigmatic illustration (right), the book contains a series of striking full-page drawings of spirits and demons with their planetary seals and signs. Each one is associated with a separate coercion spell or command. The figure in the page shown here is a very hairy Lucifer. The book is clearly concerned with treasure hunting.

RIGHT Page from the *Hollen Zwanck Schwartze Magie und Kunstz und Wunder Buch*. This image displays some familiar magical-alchemical motifs. A serpent wearing a crown has sometimes been interpreted as the cosmic spirit, while a serpent wrapped around the globe can be found, for example, in an Arabic *Book of Wonders* of the same period. The raven imagery refers clearly to the German Faust *Höllenzwang* tradition (see page 97); a variant genre of Faust manuscript was even called *The Black Raven*. The head of the Dresden Royal Library recalled how, in 1817, he received various requests for a grimoire that included an image of a raven with a ring in its beak.

Et Udtog af Cyphrianus

Jödiske Tabel.

samt

Necromantien, Demonlogien og

Göetien.

Indeholdende Matematiske, Cymie, Experimentalphysiske Konster og Widenskaber.

et

Udtog

Af den Ceremonialske Magie og Theurgien; og indeholder det som nu er her at finde:

Sortbogen

Blev först funden paa Wittenbergs Academie

Aar 1529.

i

En Marmorsteens Kiste skreven paa

Pergament.

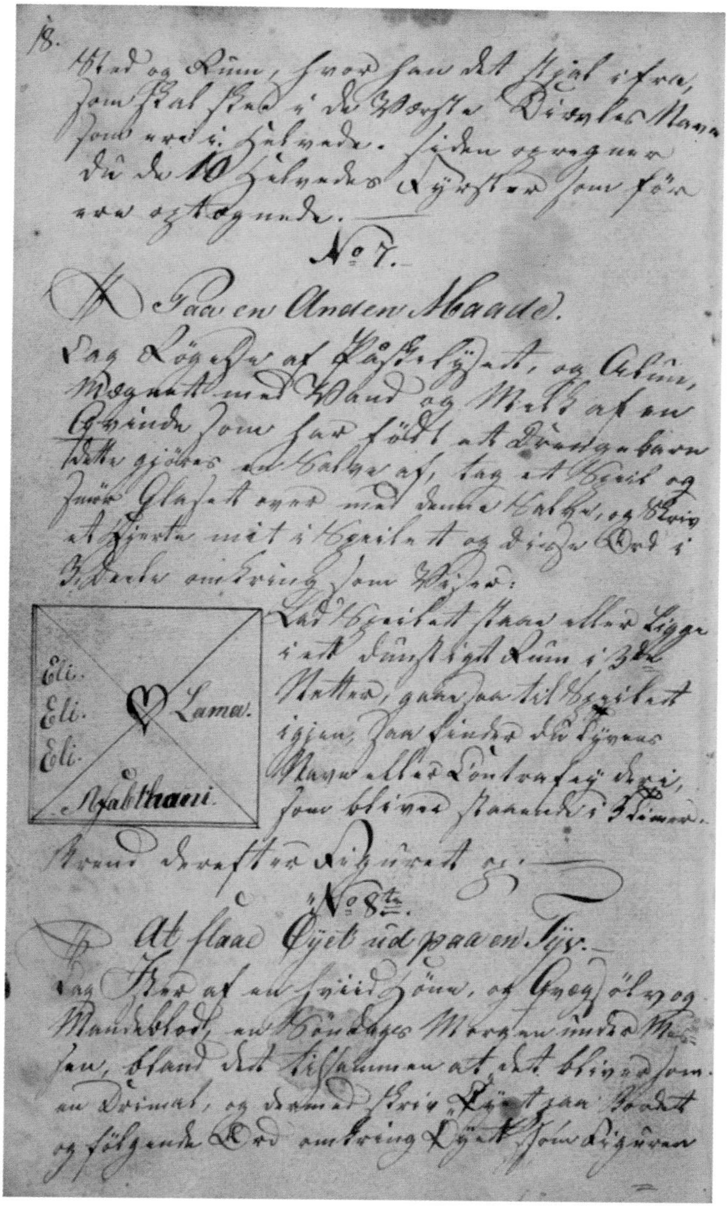

LEFT AND ABOVE Pages from a Norwegian grimoire entitled *Et Udtog af Cypherianus og Jødiske Cabal, samt Nec Cromantien, Demonlogien og Gøetien* (*A summary of Cyprianus and Jewish Cabala, as well as Necromancy, Demonology and Goetia*), on paper (c. 1800). It was obtained by the Norwegian book collector Johan Elias Schweigaard (1846–1919). The title page (left) explains that the book was first discovered in a marble chest at the Wittenberg Academy in the year 1529. This refers to a legend that such Black Books were written by Scandinavian Lutheran clergymen, who trained at a school of black arts in Wittenberg, the birthplace of Protestantism. The German influence on the contents of such books is evident from the Hebraic words *Eli, Eli, lama Afabthani* in the square diagram with the heart (above), which was part of a spell to identify a thief. This phrasing and spelling can be found in German Bibles in Matthew 27: 46, where Jesus on the cross cries out, "My God, my God, why hast thou forsaken me?"

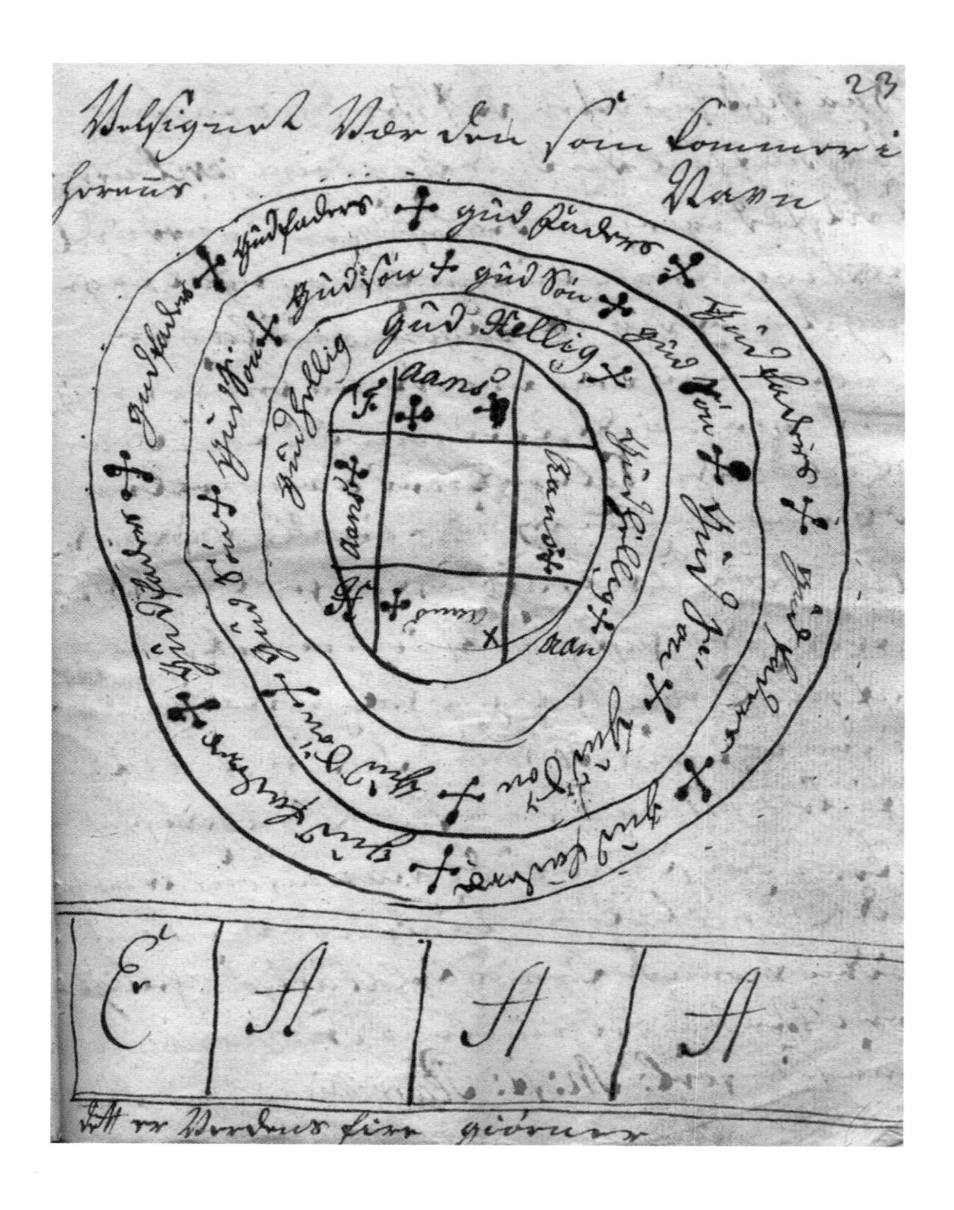

ABOVE AND RIGHT Pages from J. Olsen's *Svartebok* (*Black Book*), Stavanger, Norway (19th century). At first sight, one could almost be looking at a poor quality, late-medieval magical parchment manuscript from Western Europe, although the seal above is rather debased and contains few of the holy names, characters, and symbols found in Solomonic circle diagrams. The diminishing "Auratabul" word charm (right) can also be found in other Norwegian Black Books, and was to be written above the door of a house that had been burgled to force the thief to reappear and so be caught.

Nu Gaa: Nu Kom: Nu Lob:

Beselbus: Venere: ad: Mundus: —

...det saa skrefvet, paa Bordet en Time
...r efter, hvis noget heeld Creatur: lidt
...r af: strax kommer Trold qvinden, og
...r begierende Salt eller Mad, saa til
...efter Dammer. —

3die At een Tyv ikke skal komme
af Stæd. —

Skriv denne for manning 3 gange over
Døren som Tyven udgiid af Huuset saa
skal hand ikke undflye om hand var vel [illegible]
" fatt.

Auratabul +
Auratabu +
Auratab: +
Aurat: +
Aura: +
Aur: +

" Au: +
A: +
A +

Moridamem
Mionda Regon

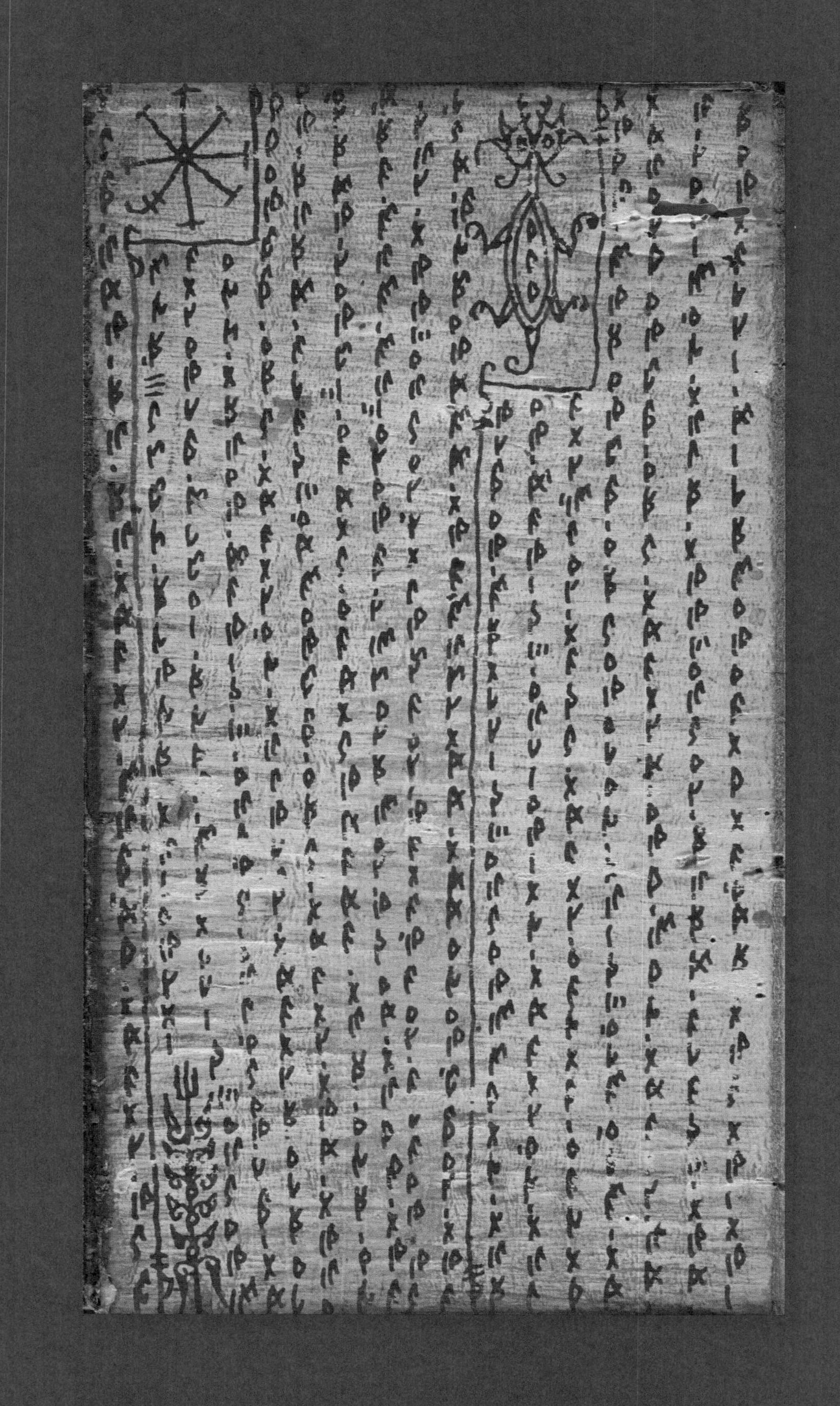

LEFT Page from a Batak *pustaha* (see page 138), a tree-bark manuscript within wooden covers, northern Sumatra (18th to 19th centuries). The image of a four-legged dragon-demon represents the mythic underworld figure of the *Naga Padoha*, who holds up the earth and was thought responsible for earthquakes. One of the divination rituals in the manuscript requires the drawing of such an image on the ground. Other parts of this *pustaha* are concerned with astrological divination, including on matters of war, and spells to protect the person and the village. As with other magic *pustaha*, the history of the transmission of the manuscript from one *datu* (magician-priest) to another is recorded.

ABOVE Batak *pustaha*, tree-bark manuscript with wooden cover decorated with plant motifs, northern Sumatra (19th century). The covers of the book can be closed and fastened tightly with plaited plant-fiber ribbons, and there is also a cord attached so that it can be carried on the shoulder by the *datu* on their journeys to clients. The contents contain the usual distinctive use of red figures and motifs outlined in black ink. The script used in such *pustaha* was an ancient Batak dialect known as *poda*, which also contained words from the Malay language. Just as in other cultures explored in this book, the use of *poda* as a secret language of instruction was to ensure it was only readable by the magician-priests.

The Pennsylvania Powwow Tradition

German migration to north America has deep roots in the colonial era, when numerous small Protestant religious groups set up communities there. Others followed for mostly economic reasons, looking for cheap farmland. The largest wave of migration occurred during the mid-19th century, and by 1890 there were around 2.8 million German-born immigrants living in the United States. Maintaining the German language through manuscript and print was important to many German–American communities, and in the 1890s there were some 800 German-language newspapers and journals being printed. Among the Pennsylvania "Dutch," who are actually descended from German-speakers, who emigrated from southern Germany, Switzerland, and Alsace during the early waves of migration, there was a particularly strong tradition of illuminated manuscript art known as *fraktur*. This was used to decorate religious documents such as baptismal certificates, but also poems, family histories, and religious literature that served a protective function. Spiritual symbols were also painted on barns in this tradition. *Fraktur* was rare, but not unknown, in the cure and charm books that were passed down by practitioners of Pennsylvania powwow, or *braucherei* (the word "powwow" likely derives from the indigenous Algonquin language), a system of folk magic or ritual healing.

Numerous powwow manuscripts survive, which demonstrate through their contents the interplay between oral tradition, manuscript, and print in the continuation of traditional

RIGHT A rare example of *fraktur* (illuminated manuscript art) in a powwow manuscript (1893). It was written for Lillie Moyer, of Tulpehocken, Berks County, Pennsylvania, as a record of a charm against "wildfire," which is a folk name for the skin infection erysipelas. While this seems to be a copy in German written from memory, the charm can also be found in *The Egyptian Secrets of Albertus Magnus*.

LEFT A collection of blessings, kept in small canvas cloth bags (made in 1887). They were copied from printed collections of charms and prayers by the hex-doctor or powwow doctor, Joseph H. Hageman, of Reading, Pennsylvania. The letters "INRI" represent the Latin *Iesus Nazarenus, Rex Iudaeorum* (Jesus of Nazareth, King of the Jews). Hageman's trade in such anti-witchcraft charms was publicly exposed during a libel trial.

May. the. 18. 1893
Decem Ber. the. 8. 1892
LILLIE
MOYE
R
April. 29.
1893.

For. the. Wildfire. to. powwow:
Weldas. Faear. Weldar Brand
Fluck. and. Schmarz. garun
nan. Blut. and. Kaldar. brand
Eckum. faka. Dech. gott. dar
harr. Ba. wara. Dech. gott. dar
Est. Dar. allar. hachsta. man.
Dar, Dech. Weldasfire. Weldar
Bran. Fluck. and. Schmarz.
garunnan. Blut. And. Kal
dan. Brand. and. allan.
Schadan. Wedar fon. Der. n
Fardraeban. kan. + + +
To. you. Lillie. A. Moyer. yours
truly. friend

Germanic folk magic in America. Some powwow doctors used cypher scripts to keep their knowledge a secret from the curious. They contain a mixture of anti-witch spells, magic to ensure a gun would shoot straight or to protect from bullets, simple religious healing charms used against natural diseases, and other folk cures. There was a significant amount of copying from printed works in German, most notably a cheap book or "chapbook" called *Der lange Verborgene Freund*, or *Long Lost Friend*, written by a German immigrant named John George Hohman and published in 1820. Another, which had its origins in an 18th-century German book of charms, was *The Egyptian Secrets of Albertus Magnus*. One of the more common powwow spells against witchcraft, taken from the *Long Lost Friend*, was to write out the following word square:

SATOR
AREPO
TENET
OPERA
ROTAS

This charm dates back to antiquity, but was still alive and well in German–American folk magic. Both books also included a German-origin charm that was originally used against the nightmare or "pressing spirit," but now employed for general house protection. The powwow doctor wrote down the charm in German, sometimes backward, on bits of paper to be concealed around the home. Second-hand furniture dealers have been known to find them hidden in old couches and beds.

بسم الله الرحمن الرحيم فايدة ان تكتب هذا
من سورة الواقعة ٨ من هذا الخواتم وتعلق
عليه وتفسيره من البر والعسر والذهب و
الفضة يرزقه الله على الاكثير احتر ايعلم عدد
الا الله تعالى ولا يموت فقيرا ابدا باذن
الله تعالى

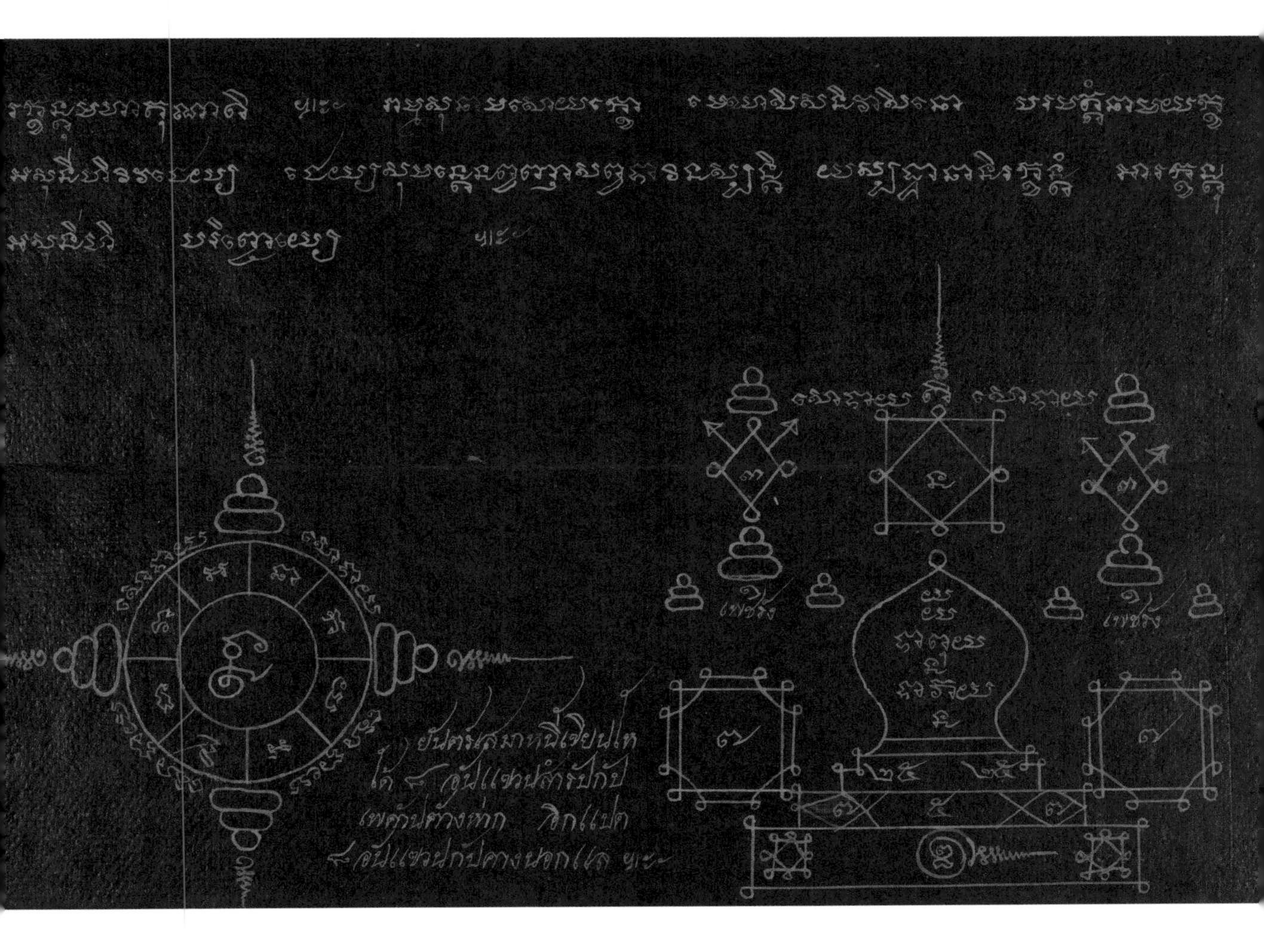

LEFT Instructions for the preparation of a magic square amulet for blessing and wealth, from a bound collection of Arabic manuscript material on paper from Kumasi, Ghana (Asante empire, mid-19th century). Among other materials in the same collection are a supplication for God to join two people in love, a diagram to be drawn on the ground to keep one safe on a journey, and a written amulet for the return of wives or slaves. A note presented to the British Museum Library along with the manuscript donation stated it was taken from Kumasi in 1874 by a "blue jacket"—the nickname for a British sailor. In that year, British forces brutally sacked, looted, and burned to the ground the Asante capital of Kumasi.

ABOVE Page from a folding book of *yantras*, written in Thai language and Khom script in yellow steatite ink on *khoi* paper (19th century). The blackening of paper for illuminated writing through the application of soot and/or lacquering the paper with plant gums, has long been practiced in Thailand and Burma. The ancient Khom, or Khoom, script was a variant of Khmer script, and was exclusive to Brahmin priests. Hence, like Sanskrit, it was used for religious, medical, astrological, and magical documents. The *yantras* in this manual were to be replicated on protective shirts, ritual metal water bowls, drums, and as protective tattoos. The Thai authorities began to phase out the use of the Khom script from the mid-19th century.

ABOVE Rolled-up Ethiopian incantation and healing scroll, written in *Ge'ez* on parchment (19th century). This scroll has a length of nearly 6.5ft (2m), and it was likely displayed in its unfurled state as a protective device in the home of the owner. This example contains prayers against witches, the evil eye, and demons—specifically one called Barya, and another called Legewon.

RIGHT Unfurled Ethiopian incantation scroll (a different example to the one above), written in *Ge'ez* on parchment, Tigray region (19th century). The image at the top left shows the arms of the four archangels bearing swords and fighting off demons. The center of this vignette is an abstract version of the eight-pointed star, formed around a cross, which represents the Seal of Solomon. The four heads on the bottom right are likely representations of the four archangels.

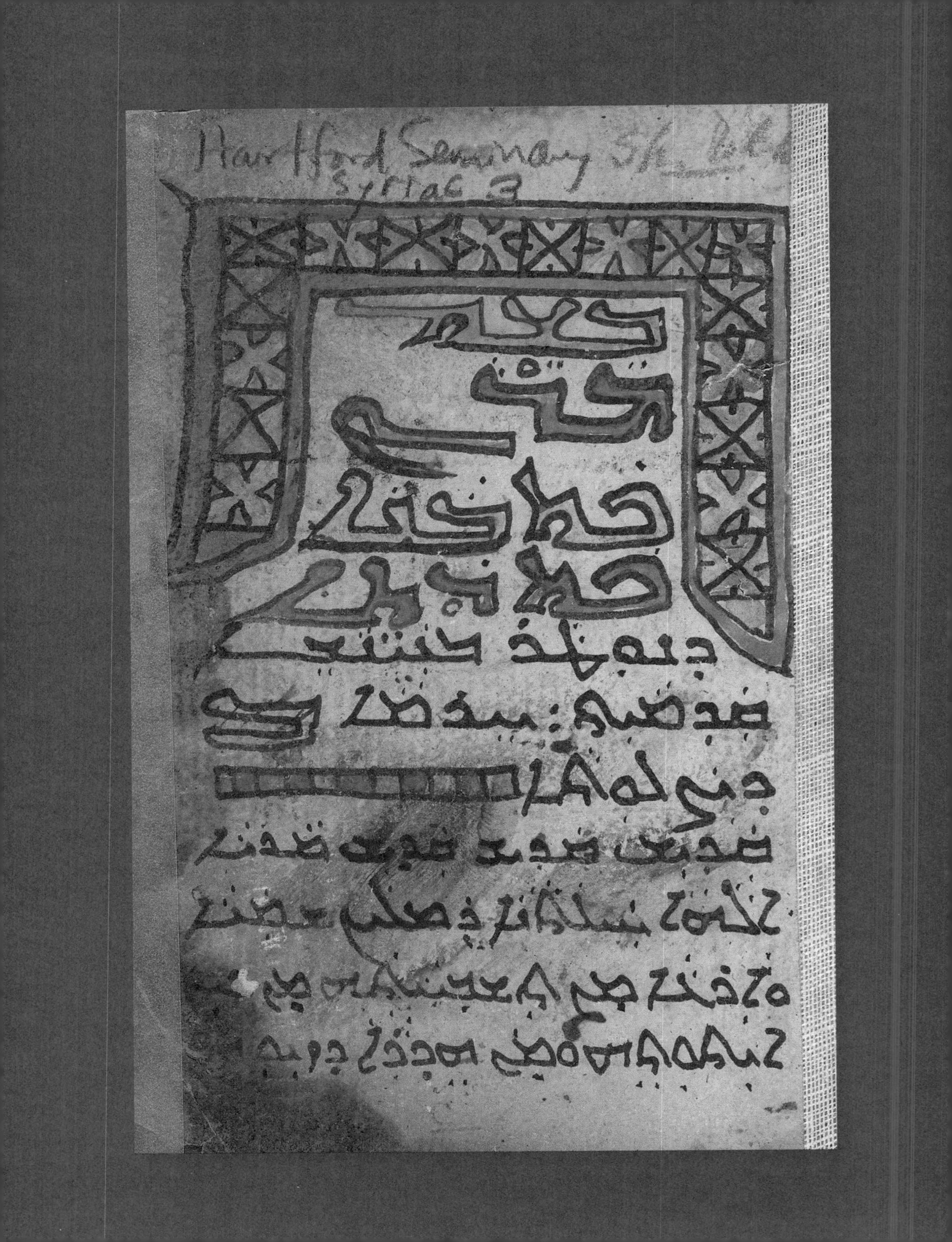

[illegible]

ܕܒܘܪܟ ܚ̈ܬܢܐ

[illegible]

LEFT AND ABOVE Images from a Christian Syriac prayer and incantation book, written in East Syriac script on paper (18th century). The manuscript consists of 80 or so pages of text and images, with illustrative depictions of saints, angels, demons, plants, animals, and various weapons. It is an example of a Syriac charm compendia, known to scholars as the "Book of Protection." As the scribe of another similar manuscript stated, it was written for the "protection of humanity from all that is evil and hostile." This tradition of astrological-magical books was particularly prominent among the Nestorian Church communities in the north of Iraq and Iran. British and American missionaries active in the region during the mid-19th century brought examples of such books back with them. The image on the right shows the four evangelists Matthew, Mark, Luke, and John, called upon in a prayer to help a man going to court.

LETTER

WRITTEN BY

LF, AND LEFT DOWN AT MAGDEBURG.

in golden
God through
n who will

py it, it shall be given; who
despiseth it, from him will
part

CHAPTER FIVE

The Power of Pulp Print

…is cursed. Therefore, …not work on Sunday, …rch; but do not adorn your …strange hair, and not carry …ll give to the poor of your …believe, that this letter is …and sent out by Christ …not act like the dumb …s in one week during which …r labors but the seventh …you shall keep holy; if you …send war, famine, pest and …nish you with many troubles …ery one, whoever he may be, …great, that you do not work …you shall regret your sins, …n you. Do not desire silver …n sensualities and desires; …de you and can destroy you.

…your neighbor is poor, feel …n, then you will fare well.

You children, honor father and mother, then you will fare well on earth. Who that doth not believe these and holds it, shall be damned and lost. I Jesus, have written this myself with my own hand; he that opposes it and scandalizes, that man shall have to expect no help from me; whoever hath the letter and does not make it known, he is cursed by the christian church, and if your sins are as large as they may be, they shall, if you have heartily regretted and repented of them, be forgiven you.

Who does not believe this, he shall die and be punished in hell, and I myself will on the last day inquire after your sins, when you will have to answer me.

And that man who carries this letter with him, and keeps it in his house, no thunder will do him any harm, and he will be safe from fire and water; and he that publishes it to mankind, will receive his reward and a joyful departure from this world.

…my command which I have sent you through …l, the true God from the heavenly throne, the …and Mary. Amen.

…OCCURRED AT MAGDEBURG, IN THE YEAR 1783.

The raging Civil War in America forced rapid innovation in the production of paper during the 1860s. Most paper production took place in the North, but a lot of the rag paper and cotton waste it relied upon came from the Confederate states. There had been a growing shortage before the conflict, due to the swelling population and burgeoning press, but the war led to the search for alternative production methods for the sake of the nation, as well as commercial interests. A fibrous substitute for rags was required urgently and various alternatives were tried, including straw, hemp, and esparto grass, but wood (which was in abundant supply) won out thanks to new industrial processes. Mechanical grinding of wood into pulp was slow, power intensive, and expensive, and so the real revolution came with a method to break down the wood chemically using sulfite acid—hence the term acid paper.

The newspaper industry was one of the big beneficiaries of wood-pulp paper, and further developments in cylindrical, high-speed printing presses during the mid-19th century, increased revenues from advertising, the spread of the railways, and the expansion of the telegraph over which copy could be wired, meant the mass-market newspaper industry took off in every city around the world, with profound cultural consequences. This was important for the proliferation of magic, as cheap newspapers provided an influential medium for advertising magic books and occult wares at home and abroad, creating a global market for popular magic in all its forms in the mail-order age. Acid paper was a poor surface for reproducing fine images but lent itself to simple reproductions and the continuation of woodcuts. The artistic qualities of pulp magic would instead come to be defined by book covers.

EUROPEAN EXPANSION OF PULP PRINT

In Europe, the French chapbook grimoires originally printed on rag paper continued to find a ready market in new pulp-paper editions, and, at the height of French colonialism, they found their way into the syncretic magical cultures of the Caribbean and Indian Ocean colonies. In 20th-century Martinique and Guadeloupe, for example,

LEFT Advertisements from the *Mystery of the Long Lost 8th, 9th and 10th Books of Moses*, New York (1948). The illustrator used fonts and imagery from pulp horror literature of the time to promote two other recent books of magic from the same street publisher (Sheldon Publications). One concerned the history of the evil eye and how to counteract its baleful influence, while the other was an edition of a much older German text called the *Long Lost Friend* (see pages 163 and 186–187).

LEFT *O verdadeiro livro de S. Cypriano* (*The true book of St Cyprian*), Lisbon (1930). This short, 16-page Portuguese *Cyprianus* contains a few *Orações* (prayers) asking God to send good spirits to intercede. It also contains a short exorcism to expel the Devil from a possessed person, as well as a simple diagram to draw on the ground to remove spells protecting buried treasure.

magical practitioners owned copies of the *Petit Albert*, *Poule Noire*, and *Dragon Rouge* (see page 112), with the latter considered a very dangerous book in local folklore. Pulp print also fostered the creation of a vibrant grimoire tradition in Spain and Portugal in the form of a new genre of *Cyprianus* magic books. Some editions of the *Libro de San Cipriano* (Spanish) and *Livro de São Cypriano* (Portuguese) tell how the book was originally brought into the world by a fabled medieval German monk and librarian named Jonas Sufurino, who was given secret knowledge after he conjured up the Devil on a mountaintop one night in the year 1001. In fact, the Iberian *Cyprianus* only dates back to the 19th century.

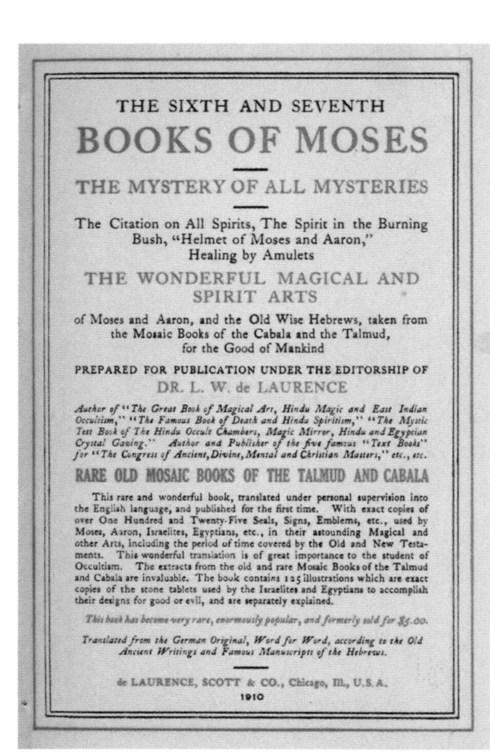

THE SIXTH AND SEVENTH

BOOKS OF MOSES

THE MYSTERY OF ALL MYSTERIES

The Citation on All Spirits, The Spirit in the Burning Bush, "Helmet of Moses and Aaron," Healing by Amulets

THE WONDERFUL MAGICAL AND SPIRIT ARTS

of Moses and Aaron, and the Old Wise Hebrews, taken from the Mosaic Books of the Cabala and the Talmud, for the Good of Mankind

PREPARED FOR PUBLICATION UNDER THE EDITORSHIP OF

DR. L. W. de LAURENCE

Author of "The Great Book of Magical Art, Hindu Magic and East Indian Occultism," "The Famous Book of Death and Hindu Spiritism," "The Mystic Test Book of The Hindu Occult Chambers, Magic Mirror, Hindu and Egyptian Crystal Gazing." Author and Publisher of the five famous "Text Books" for "The Congress of Ancient, Divine, Mental and Christian Masters," etc., etc.

RARE OLD MOSAIC BOOKS OF THE TALMUD AND CABALA

This rare and wonderful book, translated under personal supervision into the English language, and published for the first time. With exact copies of over One Hundred and Twenty-Five Seals, Signs, Emblems, etc., used by Moses, Aaron, Israelites, Egyptians, etc., in their astounding Magical and other Arts, including the period of time covered by the Old and New Testaments. This wonderful translation is of great importance to the student of Occultism. The extracts from the old and rare Mosaic Books of the Talmud and Cabala are invaluable. The book contains 125 illustrations which are exact copies of the stone tablets used by the Israelites and Egyptians to accomplish their designs for good or evil, and are separately explained.

This book has become very rare, enormously popular, and formerly sold for $5.00.

Translated from the German Original, Word for Word, according to the Old Ancient Writings and Famous Manuscripts of the Hebrews.

de LAURENCE, SCOTT & CO., Chicago, Ill., U.S.A.

1910

LEFT William Lauron De Laurence's edition of *The Sixth and Seventh Books of Moses* (1910). He rightly supposed "it would be a good seller." The cover contains all of De Laurence's usual boastful and fabricated claims, such as that the book was translated from German under his personal supervision and editorship. In fact, he had bought the printing plates from an acquaintance named Fred Drake, who published non-magical do-it-yourself manuals.

A thriving trade in popular Jewish magic books also developed in Eastern Europe around this time. Some were reprints of works that had been published in central Europe during the 17th and 18th centuries, while a new wave of books was predominantly produced by Jewish printing houses in Lviv, Warsaw, and Piotrków. The authors were often Hasidic rabbis, from the mystical, spiritual, and revivalist Hasidism movement that developed within Eastern European Judaism during the 18th century. Hasidism drew upon folk magic and Kabbalah, as well as the power of prayer. One such author was the Polish rabbi, Hayim Isaiah Halbersberg (1844–1910), who published a popular collection of textual spells and charms taken mostly from older works that were written in a mixture of Hebrew, Yiddish, and Polish. He asked his readers "to conceal the performance of the charm from people, lest people scoff. Only a believer can perform the charm and have it work." The Jewish astrologer, amulet writer, and Kabbalist Abraham Hamuy (1838–1886), who was born in Aleppo, was another major influence, particularly in the Jewish communities of the Ottoman Empire and in Spain. He published seven or so books on magic, charms, and divination—some printed in Izmir, Turkey—that were distributed widely and influenced the East European print tradition.

Among such a scattered and oft-persecuted diaspora, particularly at a time when millions of East European Jews were being displaced, this new wave of popular print meant that Jewish people in search of charms and talismans no longer needed to rely solely on personal access to Kabbalah experts and rabbis. Over 2.5 million East European Jews arrived in the United States between the 1880s and 1920s to escape the Russian pogroms, ethnic tensions in the Austro-Hungarian Empire, and economic hardship. No doubt some print magic books found their way with them for protection and health on the journey, and were used in their new homes. For later arrivals on American shores, though, there was already the *Sixth and Seventh Books of Moses*—a cheap grimoire that purported to draw upon venerable Jewish magical tradition—thanks to an earlier wave of German emigration in the 19th century. Said to divulge the secret wisdom of Moses beyond the five books that make up the Torah, or Pentateuch, it contained numerous images of talismanic seals and pseudo-Hebraic occult words for protection and health. Among the long-established Pennsylvania Germans, the *Sixth and Seventh Books of Moses* had an ambivalent reputation. Stories circulated of people bewitched by hex doctors in possession of a copy, as they could use it to both curse and cure. Yet, the thirst for such secret "Biblical" magic spread far beyond the German–American population. Thanks to occult publishing entrepreneurs, such as William Lauron De Laurence in Chicago, who supplied books by mail order, the *Sixth and Seventh Books of Moses* became available to an international readership, and was adopted as a sacred text in the Caribbean by Rastafari and by practitioners of *obeah*, a magico-religious healing tradition based on West African spiritual traditions.

The multicultural street culture of New York during the 1940s generated a raft of new pulp magical works that were cheaper and lower quality than De Laurence publications. Some had spines that had been stapled rather than glued, and looked as if they were products of early photostatic printing with terrible image quality. Titles included such works as *Terrors of the Evil Eye Exposed* and *The Master Key to Occult Secrets*, written by mysterious adepts with French-sounding names. Some of the better-quality productions were by the Dorene Publishing Company, which was founded by Joseph Spitalnick, the son of Russian immigrants. These publications found a ready audience among the Africana esoteric diaspora, especially after the British colonial authorities banned De Laurence publications from Jamaica in 1940. In 1949, a British anthropologist also found Dorene magic titles circulating in Belize.

POST-COLONIAL PULP PRINT EXPLOSION

Spanish and Portuguese manuscript grimoires arrived in Central and South America from the early decades of Hispanic conquest in the 16th century, where the colonial Inquisitions made it their business to try and suppress their spread in the New World. But the new age of pulp magic books in the post-colonial era had a big influence on the magical milieus in the region. In colonial Brazil, no printing presses were allowed until 1808, but once this edict was relaxed presses were set up in Rio de Janeiro and elsewhere, and within a few decades a distinctive publishing industry began to develop. From around 1860, a genre of chapbook publishing known as *folhetos*, or *literatura de cordel* flourished, particularly in northeastern Brazil. These

RIGHT *La Clavicula del Hechicero o El Gran Libro de San Cipriano* (*The Sorcerer's Key or The Great Book of St Ciprian*), Buenos Aires (1970). This grimoire was purportedly translated and edited by the "celebrated" but mysterious Dr. Moorne. We first hear of him in the early 20th century as the author of several Spanish magic books that sold for a few pesetas, but he subsequently became a legendary occult figure in Argentina.

LEFT Advertisements from Lewis de Claremont's *The Ancient's Book of Magic, New York* (1940). De Claremont was a pseudonym. The advert shows how the publisher, Dorene, successfully tapped into the Latino interest in magic books, both in New York and in Central America, by producing Spanish-language editions (see pages 204–205).

books featured distinctive woodcut images, and prose and poetry concerned with tales of bandits, romance, and religion intended for the masses. The term "*cordel*" denotes the strings that held such chapbooks suspended, when sold at markets or in the streets. There are no examples of early chapbook grimoires in this specific genre, but the *folhetos* proved there was a market for literature among the poor and from the 1870s, cheap magic books began to appear in other formats. The first-known edition of the *Livro de São Cypriano* (see page 173) in Brazil was advertised in a Rio de Janeiro newspaper in 1876, though it may have been a Portuguese import.

The entrepreneurial bookseller, Pedro da Silva Quaresma, was a major influence on the *Livro de São Cypriano*'s increasing popularity in Brazil during the late 19th and early 20th centuries. He produced comprehensive catalogs of his wares, including the *São Cypriano*, and made good use of newspaper advertisements for sales promotion. According to a Brazilian news item from 1916, the best-selling books that September were, indeed, the *Livro de São Cypriano*, with 1223 copies sold, and a cheap divination manual, called the *Livro da Bruxa* (*Book of the Witch*), with 1419 copies. Journalistic comments on the *Livro de São Cypriano*'s popularity can be found during the 1930s and 1940s, followed by a significant rise in the number of distinctly Brazilian editions in the 1950s and 1960s. This was fuelled by the growing market for manuals concerning the Afro-Brazilian syncretic religions known as *candomblé* and *umbanda*, which blended Catholicism with African religions, spiritism, and indigenous beliefs.

Mexico was the pulp publishing powerhouse of Central America, and books from its presses circulated down into South America. As a consequence, pulp grimoires also seeped into the beliefs and practices of indigenous peoples, as well as those of Spanish and Portuguese extraction. Among the Nahua population of Central Mexico, the *Libro de San Cipriano* was considered a dangerous book that would make you crazy, if you were not sufficiently powerful. It was thought it could be used to transform oneself into animals. A study of folk medicine and magic among the indigenous Inga of Colombia shows how *curenderos* (folk healers) mixed traditional medicinal plants with Catholic devotional objects and pulp occult works, including the use of the *Libro de San Cipriano* and titles such as *Magia Roja* and *Magia Verde* published in Mexico City. The Inga considered Germans as having a particularly profound knowledge, because of the story of Jonas Sufurino told in the *Cipriano* (see page 173), and so they also prized "white magic" pulp books attributed to the German physician Johann Gaspar Spurzheim (1776–1832), who was one of the leading proponents of the pseudo-science of phrenology. The legend of St Cyprian was also a particular preoccupation with the Quecha storytellers of southern Peru. The Quecha complained that Catholic priests had taught the *nak'aq* (gringos and outsiders, believed to be harvesting human fat from the Quecha) to read and write, enabling them to access the power of the *Libro de San Cipriano* while refusing to spread literacy in their own communities. The Quecha believed that their lack of access to the book put them at a disadvantage, and left them unable to use it for protection from the *nak'aq*.

THE EMERGENCE OF COLOR PRINT

The 1940s and 1950s saw a revolution in pulp literature in the United States, when inexpensive color-printing techniques enabled the creation of cheap, glossy, and eye-popping color covers that belied the black and white, coarse acid paper within. While publishing houses, such as Penguin, in the UK stuck to staid text-based covers for their paperback editions, the American covers for pulp science fiction, westerns, crime novels, and romances featured garishly dramatic scenes and salacious sexual imagery, representing the sensational stories that lay within. This was perfect for a new era of pulp grimoires, and it was the street publishers of Central and South America that exploited it to the full. We know little of the artists who designed these covers, but they displayed a distinctive creative energy, drawing on a blend of Western filmic and folkloric imagery, stereotypes of the mystical orient, and notions of indigenous know-how.

The pulp revolution was a global phenomenon that influenced every continent and many cultures. In Republican China (1912–1949),

RIGHT *La Magia Negra y Arte Adivinatoria* (*Black Magic and Divinatory Art*), Buenos Aires. Despite its striking cover of a stylized indigenous person, the content of this book is a mishmash of basic information on European magic and divination. Those seeking instructions for black magic and conjurations would have been disappointed. There is a section on cartomancy, and a crude set of talismans, including the ubiquitous diminishing abracadabra.

LEFT *La Magia Amorosa o Verde y la Magia Roja* (*The Magic of Love and Green and Red Magic*), Buenos Aires (1969). This is presented as a companion to La Magia Negra (see page 177). Both were published by Caymi, which was founded in the 1940s to print and sell cheap, pulp literature across Argentina. It also published gaucho cowboy novels, martial arts manuals, and medical and sex advice.

a flourishing market for cheap print occult literature developed. This included the promotion of *shushu*, or "calculations and arts," which were once the preserve of a scribal-priestly elite, and *xinling kexue* (spiritual science), represented by a raft of popular publications on yoga, astrology, geomancy, and meditation. Just as the new print industry transformed popular engagement with Daoist and Buddhist religion in China, so it also re-shaped the world of folk magic. Shanghai became the center of this industry, and publishers there placed thousands of advertisements for occult services and books in the newspapers. A publishing house known as The Press, founded by Wu Lugong in 1917, was at the forefront, publishing numerous cheap books of *qishu*, or "occult arts," in vernacular Chinese. Perhaps the most influential publication of the period was the *Compendium of Secret Talismans* by Yu Zhefu, published by the Chenzhou Spiritual Society, which contained talismans for the usual range of purposes, including healing, exorcism, invoking spirits, and finding lost or stolen property. The *Compendium* was first published in several editions as a reproduced hand-written manuscript, rather than a typescript, using recent photolithographic printing techniques. This was not for cost reasons, rather the aim was to present the manuscript authenticity of Yu Zhefu's hereditary knowledge of talismans from his Chenzhou ancestors, a city famed for its talisman culture. Like De Laurence in America, Yu Zhefu took full entrepreneurial advantage of the tools of modernity, advertising in the press, setting up a Research Institute of Chenzhou Talismans, offering free membership to purchasers of his *Compendium*, and providing distance-learning correspondence courses.

Popular printed books of magic and charms in South Asia date back to the 1860s, with an anonymous 12-page collection of exorcisms and treatments called the *Bangiya Biswasya Mantrabali* (*Trusted Bengali Mantras*) published in 1863. Over the decades, other such publications, which appealed to both Hindus and Muslims, were produced that concentrated on love magic and healing spells. Then, in the mid-20th century, their production style shifted to larger 200-page books that were often printed in red ink, undated, and anonymously written to give a sense of mystique and agelessness. Folk magic rituals were mixed with Hindu mantras, magic word squares, and passages from the Qur'an. This emerging genre was also defined by its contents, which showed a new market for pulp magic among shopkeepers and managers of small businesses across South Asia. Much larger volumes developed in the 1970s and 1980s, with lively full-color covers representing deities and occult symbols. They sold widely in bazaars, train stations, bus depots, and road-side stalls. The various spells and rituals to increase sales and hinder rivals are rich and diverse. A book published in Assam advised failing business owners to change their fortunes by filling up a glass bottle with mustard oil and throwing it into a lake or river. One publication had the following charm for owners whose employees refused to work: "On a Saturday quietly pick up any piece of iron that you find on the way to your factory or workplace. Once you arrive at your workplace wash the iron nail with buffalo's urine and then wash it with Ganga water. Thereafter, hammer it into the wall in the room where your employee works." Another manual detailed how to destroy the spinach of competing spinach farmers and damage the clothes being cleaned by rival washerwomen.

LEFT *Brihat Indrajala* (*The Big Art of Magic*), published in Hindi. This is one of numerous such magic books with sensational covers published in Hindi and Bengali from the 1970s onward. They were mostly published in Kolkata or Dhaka. Skulls were prominent on the Hindu covers, sometimes with the depiction of an Aghori guru with a human skull, and the great goddess Kali, with her tongue sticking out. Here, we see Shiva with a crescent moon on his headdress, holding his trident.

O Pequeno Livro de
São Cipriano

LEFT *O Pequeno Livro de São Cipriano* (*The Little Book of Saint Cyprian*), São Paulo. In keeping with the cigar-munching demon on the cover, a significant portion of this 64-page book is concerned with countering evil spirits through *Orações* (prayers), exorcisms, and rituals. It is distinctively Brazilian, in that it also contains instructions for countering *macumbeiros*—the practitioners of the Afro-Brazilian spirit religion known as *macumba*. One of the prayers against enemies appeals to St Gabriel to help: "I know they plot against me through the nefarious help of macumbeiros, who seek to overthrow me."

ABOVE *Antigo Livro de São Cipriano, O Gigante e Verdadeiro Capa de Aço* (*The Ancient Book of St Cyprian, or the Giant and True Protection of Steel*), Rio de Janeiro. With over 500 pages, this is the largest book in the South American *Cyprianus* genre, and one of the most recent iterations, which has been through at least 30 editions since the 1970s. It contains over 100 *Orações* (prayers), ranging from the cure of cancers to protection from demonic persecution, alongside sections on cartomancy and astrological signs. The mysterious editor, N.A. Molina, was also the supposed author of other occult books published in the 1970s, including one on the power of *quimbanda*—a variant of *macumba* (see left).

ABOVE *Kardoné el mago: los grandes misterios y la magia del amor. Oraciones y conjuros de gran poder* (*Kardoné the magician: the great mysteries and the magic of love. Prayers and spells of great power*), Medellín, Colombia. This pulp magic book of 35 pages cost 15 pesos and seems to have been a promotional tool for a mail-order occultist named "Professor Kardoné," who advertised that he took on "all kinds of inquiries by correspondence." The book references Cyprianus and Albert the Great, and contains various spells for sexual attraction and against harmful magic. One love ritual instructs: "In order to be loved by all women, whoever wishes to do so must carry a small green silk bag containing the heart of a dove and the eyes of a cat so that it touches the skin of the left breast."

RIGHT *Misterios de las ciencias ocultas antiguas y novísimas teorías sobre la magia blanca y magia negra* (*Mysteries of the ancient occult sciences and the latest theories on white magic and black magic*), Madrid (1923). This is a wordy, popular book about magic rather than a manual, which contains no images other than the distinctive cover. It was written by Ciro Bayo (1859–1939), a Spanish writer, translator, and traveler, who authored another book on occultism a few years later.

MISTERIOS
DE LAS
CIENCIAS OCULTAS
MAGIA BLANCA MAGIA NEGRA
POR EL
RAJAH SACANDALA
EDITORIAL
RUBIÑOS
Riquer

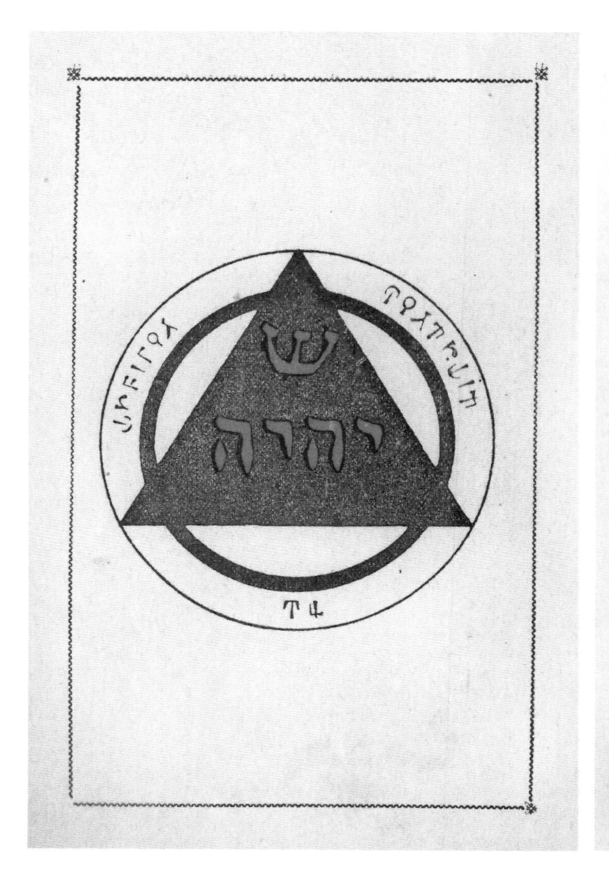

ABOVE Colored plates from *Le livre secret des grands exorcisms et bénédictions* (*The Secret Book of Great Exorcisms and Blessings*) by Abbé Julio, Vincennes, Paris (1908). As well as dozens of protective and healing prayers, often instructed to be written on parchment, this 613-page book contains a series of colored talismans, which are not always related to the text but are meant to impress the reader. The image of the dragon with pseudo-script relates to an exorcism for evil spirits and demons. It involves a dialogue between Saint Marguerite and a dragon, which she condemns as a "wicked spirit of disorder." Abbé Julio's real name was Julien-Ernest Houssay (1844–1912), and he was an ordained priest in France, who left the Catholic Church under a cloud and subsequently tapped into the world of popular magic with his protective prayers, pentacles, and talismans. His publications were a 20th-century revamp of the old 18th-century French print grimoires.

RIGHT *Albertus Magnus: Being the Approved, Verified, Sympathetic and Natural Egyptian Secrets.* It is stated in the preface that "this book was issued, in order to bridle and check the doings of the Devil." The book consists of a large collection of medical recipes, magical cures, and spells. It is not a book of conjuration, but there are a number of simple rituals for countering witchcraft, including one entitled "To Cause a Witch to Die within One Minute." The front page of the book declares that it was translated from German, and this edition of *Albertus Magnus* was, indeed, based on an earlier German chapbook of cures and spells circulating in America, known as the *Egyptische Geheimnisse* (*Egyptian Secrets*).

PRICE $1.00.

ALBERTUS MAGNUS

Being the Approved, Verified, Sympathetic and Natural

EGYPTIAN SECRETS

—OR—

White and Black Art for Man and Beast.

REVEALING THE

Forbidden Knowledge and Mysteries of Ancient Philosophers.

Der

Lang Verborgene Freund,

enthaltend

Wunderbare und Probmäßige

Mittel und Künste

für

Menschen und Vieh.

Herausgegeben von

Johann Georg Hohmann.

Demselben ist beigefügt

Dr. G. F. Helfenstein's

vielfältig erprobter

Hausschatz der Sympathie.

Evangelium Marci, Cap. 11, v. 22. 23.

Harrisburg, Pa.

Gedruckt bei Scheffer und Beck,

1853.

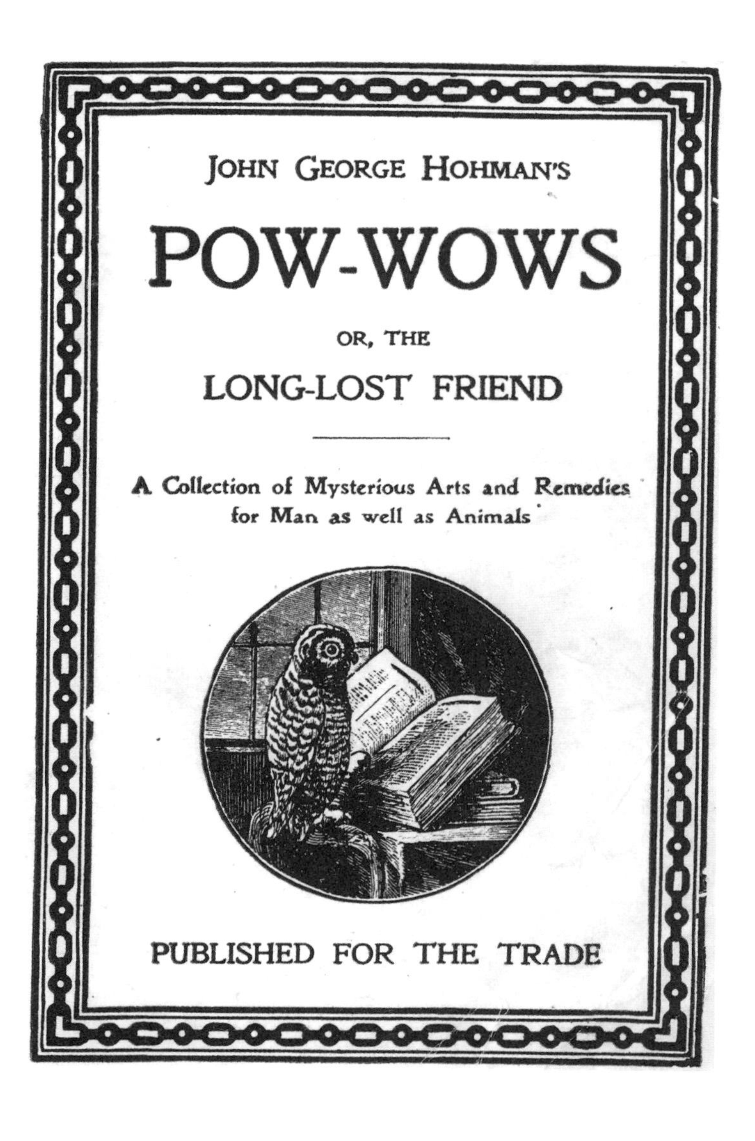

LEFT AND ABOVE The *Lang Verborgene Freund*, or *Long Lost Friend*, was one of the most influential magic books among Pennsylvania Germans. The first German edition was published in America in 1820 by an immigrant named John George Hohman. He produced an English version in 1846, and a more polished translation by another hand was published in Harrisburg, Pennsylvania, in 1856. The latter was reprinted many times. Some later editions were renamed *Pow-Wows* (see above). Much of its contents derived from an 18th-century text circulating in Germany called the ***Romanusbüchlein***, which purported to be a compilation of Romani folk magic. The cures and spells in the *Long Lost Friend* were similar to those in Albertus Magnus (see page 185), and included charms to protect against thieves, witches, guns, and house fires. To win at cards every time, it instructed: "Tie the heart of a bat with a red silken string to the right arm."

Letters from Heaven

Known as Saviour's Letters in Britain, *Himmelsbrief* in Germany, and *Lettres du ciel* in France, these apocryphal holy messages from the skies circulated during the early medieval period. They claimed to be copies of letters sent to Earth by Christ for the benefit of humankind, and found under rocks in various places and times. Such celestial letters were already printed as popular literature in the 18th century. They are a classic example of the interplay between manuscript and print, as the printed letters usually demanded that the contents be copied out by hand and passed on just like a chain letter. Folklorists in the 19th century found many such handwritten copies kept in people's homes. Most cheap print examples were in black and white, but publishers also produced simple colorized versions. While they were often folded and worn for protection, the poor would also paste them on the walls of their cottages, and so the color versions were both ornamental signs of piety and had talismanic properties.

In Germany, *Himmelsbrief* were an aspect of a broader German *schutzbrief* (protection letter) tradition, which consisted of a single printed sheet that included a mixture of prayers, religious images, quotes from pertinent Bible passages, and magical formulae tailored to the protection of the individual or the home. There was also a sub-genre of *kugelsegen*, or "bullet blessings" to prevent the recipient from being shot. Many examples of these were found on dead Prussian soldiers on the battlefields of the Austro-Prussian War of 1866 and the Franco-Prussian War of 1870–1871. The First World War heralded a new resurgence of *Himmelsbrief* printing, causing some German pastors to criticize the popularity of such "superstition." In one sermon a pastor told his flock, "Reliance on *himmelsbriefe* is folly and sin, but some of the warnings that are at the end inviting you to live as pious, honest German soldiers are good and useful." Among the Pennsylvania-German community, there was also a martial market. In 1918, one year after the United States entered the First World War, the Aurand publishing company of Beaver Springs,

Himmels-Brief,

mit goldenen Buchstaben geschrieben, und ist zu sehen in der Michaelis-Kirche zu St. Germain, wird genannt Gredoria, wo der Brief über der Taufe schwebt. Wer ihn angreifen will, von dem weichet er, wer ihn aber abschreiben will, zu dem neiget er sich und thut sich selbst auf.

LEFT Example of a *Himmelsbrief*, hand-colored lithograph, printed in Reutlingen, Germany (19th century). This heavenly letter depicts the common image of an angelic herald bearing a palm leaf. It begins with a command not to work on Sundays, and includes a prayer for protection from fire and flood. It claims the original letter is written in golden letters in the Michaelis church at St Germain.

Copy of a Letter Written by Our Bleſſed Lord and Saviour JESUS CHRIST,

AND found eighteen miles from Iconium, ſixty-three Years after our Bleſſed Saviour's Crucifixion.

Tranſmitted from the Holy City by a converted Jew.

Faithfully tranſlated from the original Hebrew copy, now in the poſſeſſion of the Lady Cuba's family in Meſopotamia.

This Letter was written by JESUS CHRIST and found under a great ſtone, round and large, at the [illegible] of the Croſs. Upon the ſtone was engraven—"BLESSED IS HE THAT SHALL TURN ME OVER." All people that ſaw it, pray to God earneſtly, and deſired, that he would make this writing known unto them; and that they might not attempt in vain to turn it over. In the mean time there came out a little child, about ſix or ſeven years of age, and turned it over without aſſiſtance, to the admiration of every perſon who was ſtanding by. It was carried to the city of Iconium, and there publiſhed by a perſon belonging to the Lady Cuba.

On the letter was written,

THE COMMANDMENTS OF JESUS CHRIST,

Signed by the Angel Gabriel, ſeventy-four years after Our Saviour's birth.

FAIRER THEN THE CHILDREN OF MEN PSAL XLV.

King Agbarus's Letter to our Saviour, and Our Saviour's Anſwer. Alſo, His Cures and Miracles. Likewiſe, Lentulus's Epiſtle to the Senate of Rome, containing A Deſcription of the Perſon of Jeſus Chriſt.

A LETTER OF JESUS CHRIST. CHRIST's Cures and Miracles.

ABOVE Example of a Saviour's Letter, hand-colored engraving, England (18th century). This letter, which claims to be "faithfully translated from the original Hebrew copy," relates to the apocryphal correspondence between Jesus and King Abgar, or Abgarus, V, which has its origin in the 4th century CE. The letters were supposedly exchanged in the last year of Jesus's life.

Pennsylvania, was commissioned by a private client to print a batch of *Himmelsbrief* that would be offered as protection to the state's draftees and members of the National Guard. The Aurand family were Calvinists and the mainstay of their company was the publication of Bibles for nationwide distribution.

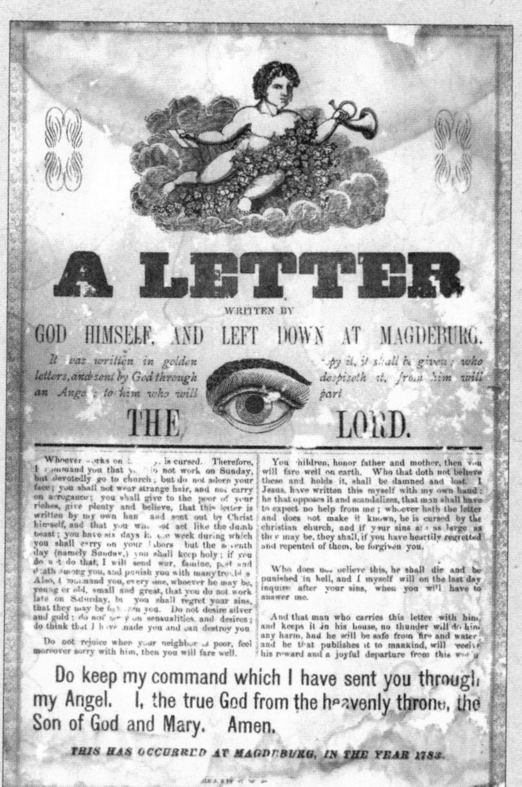

A LETTER

WRITTEN BY

GOD HIMSELF, AND LEFT DOWN AT MAGDEBURG.

It was written in golden letters, and sent by God through an Angel; to him who will [illegible] *copy it, it shall be given; who despiseth it, from him will part*

THE LORD.

Whoever works on [illegible], is cursed. Therefore, I command you that you do not work on Sunday, but devotedly go to church; but do not adorn your face; you shall not wear strange hair, and not carry on arrogance; you shall give to the poor of your riches, give plenty and believe, that this letter is written by my own hand and sent out by Christ himself, and that you will not act like the dumb beast; you have six days in the week during which you shall carry on your labors but the seventh day (namely Sunday,) you shall keep holy; if you do not do that, I will send war, famine, pest and death among you, and punish you with many troubles. Also, I command you, every one, whoever he may be, young or old, small and great, that you do not work late on Saturday, but you shall regret your sins, that they may be forgiven you. Do not desire silver and gold; do not [illegible] on sensualities and desires; do think that I have made you and can destroy you.

Do not rejoice when your neighbor is poor, feel moreover sorry with him, then you will fare well.

You children, honor father and mother, then you will fare well on earth. Who that doth not believe these and holds it, shall be damned and lost. I Jesus, have written this myself with my own hand; he that opposes it and scandalizes, that man shall have to expect no help from me; whoever hath the letter and does not make it known, he is cursed by the christian church, and if your sins are as large as they may be, they shall, if you have heartily regretted and repented of them, be forgiven you.

Who does not believe this, he shall die and be punished in hell, and I myself will on the last day inquire after your sins, when you will have to answer me.

And that man who carries this letter with him, and keeps it in his house, no thunder will do him any harm, and he will be safe from fire and water; and he that publishes it to mankind, will receive his reward and a joyful departure from this world.

Do keep my command which I have sent you through my Angel. I, the true God from the heavenly throne, the Son of God and Mary. Amen.

THIS HAS OCCURRED AT MAGDEBURG, IN THE YEAR 1783.

RIGHT "A Letter Written by God Himself." This example of a *Himmelsbrief*, which begins with a curse against those who do not keep the Sabbath, is known as the Magdeburg Letter. It was the most popular genre of heavenly letter among German Americans. The letter supposedly fell from the sky in 1783, yet this miraculous appearance is spurious, as versions were being printed in Germany as far back as the 16th century.

219

(3. *Innen*)

(IN)

ADOYAH)
(CHANANYAH
AHMEN)
GOTTES

DES VATERS (+) GOTTES + DES + SOHN
GOTTES + DES + HEILIGEN + GEISTE
(YES CHAYAHETH RACHMYEL)
DAS

Unschuldige heilige Blut JESU CHRISTI des Sohnes Gottes m
en Sünden und gebe euch Geister die ewige Ruhe und

JESUM CHRISTUM (+) GOTTES SO
(AWYEL)

Also erlöse euch Geister JESUS CHRISTUS von aller Qual und
Schätze die dahier verborgen seyn durch das vergoss

EEL ELYON (JESU CHRISTI) EHE
AMEN

(OPTIMO SUCCESSU REMISSUM.)

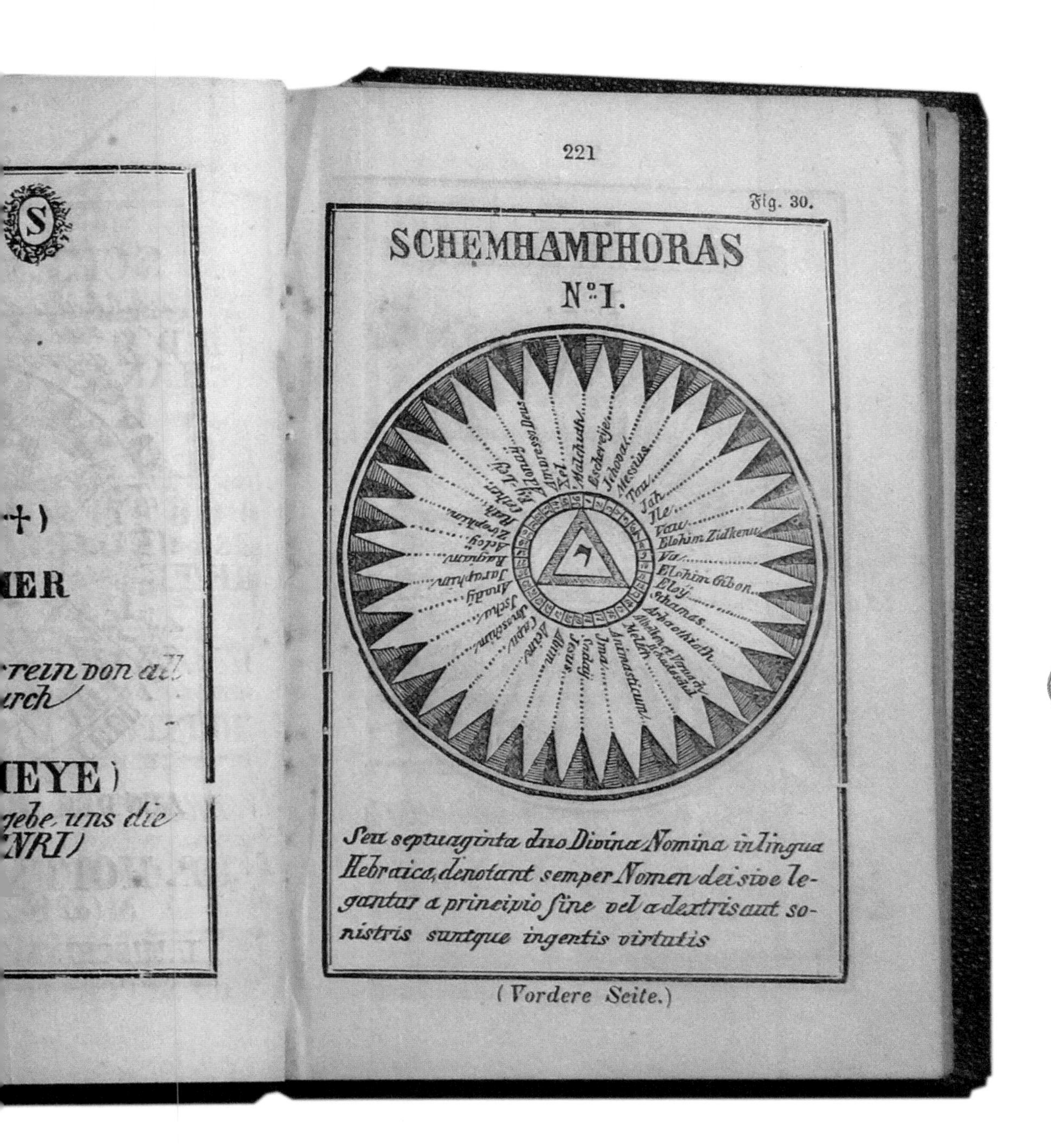

ABOVE A fold-out page from a German language edition of the *Sechste und Siebente Buch Mosis* (*Sixth and Seventh Books of Moses*). This edition was a compilation from various German grimoire manuscripts in circulation during the 18th and early 19th centuries, which was put together and published in 1849 by the Stuttgart antiquarian bookseller Johann Scheibel. The first German–American edition was printed in 1865, and sold for $1.50. It was produced by the New York book dealer, Wilhelm Radde, who also sold copies of the *Egyptische Geheimnisse* (see page 185). *Schemhamphoras* is a Kabbalistic name for God, and also the title of a Kabbalistic work printed in Germany in 1686, which was included in the Radde edition. The round diagram is two sided and denotes the power of God. It lists the 72 Hebrew divine names for the Lord (36 on the side shown, and 36 on the other side).

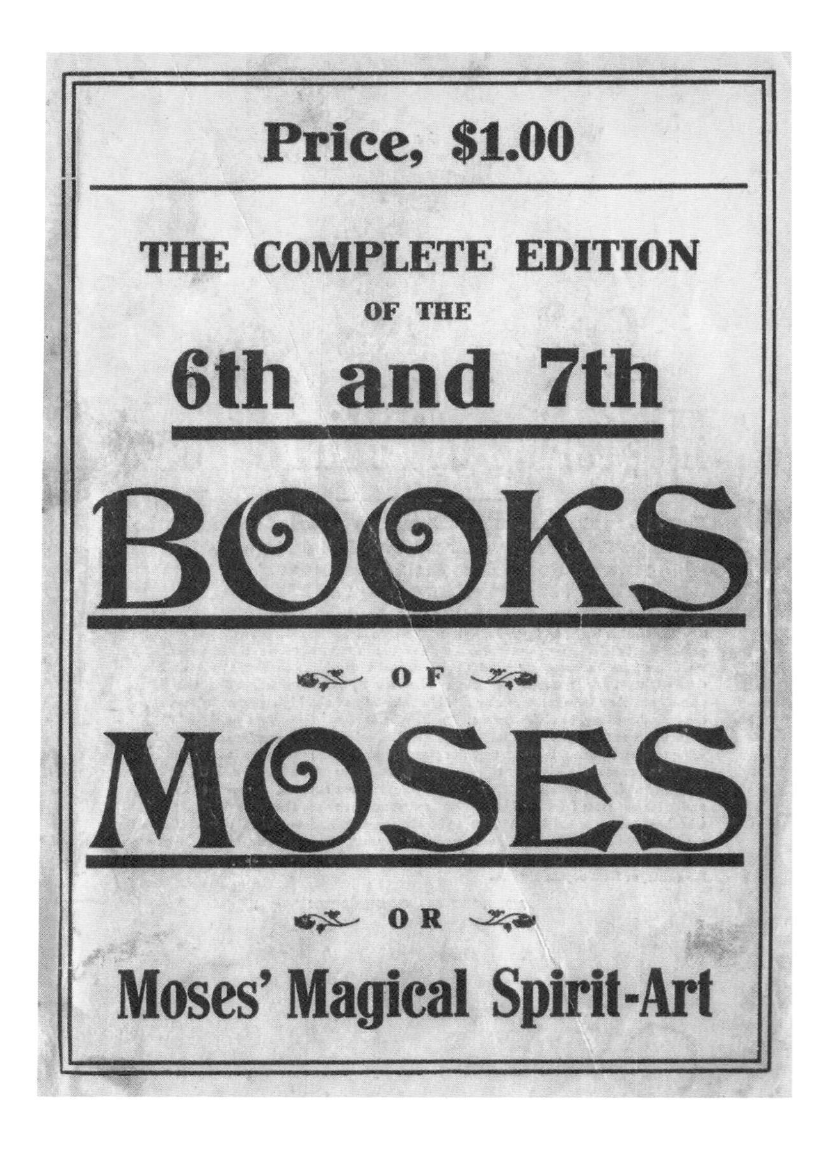

ABOVE The first English print version of the *Sixth and Seventh Books of Moses*, translated from the Radde edition, was published in New York in 1880. The above cover is from one of the many thousands of paperback copies printed during the early 20th century, using near identical printing plates. Various poor-quality advertisements for other pulp books were included, such as *Clog Dancing Made Easy* and *The Book of Great Secrets*. There is no date or publisher given in these street editions, just the statement "Published for the Trade."

RIGHT One of the numerous crudely engraved images of talismans, characters, and seals from a paperback edition of the *Sixth and Seventh Books of Moses*. There included seals to conjure the "Power Angels" and "obedient Angels," seals to place on the ground to make spirits give up their buried treasures, and one called "Balaam's Sorcery" to bring vengeance upon one's enemies. These images were full of pseudo-Hebraic characters.

DIAGRAM

ILLUSTRATING THE

SYMBOLS EMPLOYED BY THE ISRAELITES

IN THEIR

LAWS OF MAGIC.

THE GREAT BOOK OF MAGICAL ART, HINDOO MAGIC AND INDIAN OCCULTISM

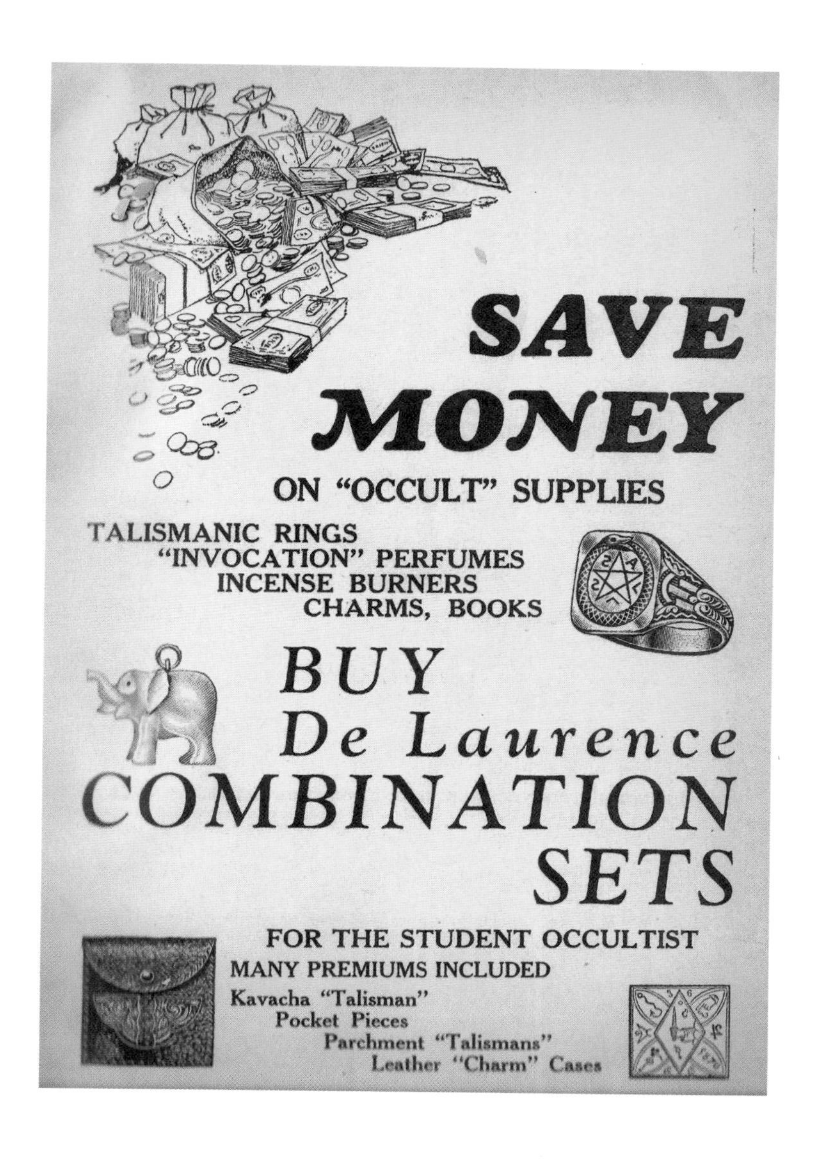

LEFT *The Great Book of Magical Art, Hindoo Magic and Indian Occultism*, Chicago (1914). The title does not reflect the content of this book, which was basically a plagiarized edition of Francis Barrett's *The Magus* from 1810 (see pages 124–125). Among its 323 pages, there is a series of unattributed engravings and photos showing scenes from India, portraits of early modern occultists, and adverts for De Laurence's magic merchandise. He rather shamelessly included an "important notice" stating that his book was protected by copyright. De Laurence also wrote in the preface that "the cost of publishing a work of this kind is very great, there is little or no profit." While it certainly was not a cheap product, with a cover price of $15, it appears to have sold quite well, and was even found in bookshops in Ghana.

ABOVE A page from William Lauron De Laurence's publishing company *Catalog* (see pages 200–201). By the 1930s, this annual *Catalog* was running to well over 500 pages of products. The range on show is quite dizzying, from the "world-renowned" kavacha talisman which "destroys all dangers and difficulties" to his special "temple incense," and his unique Hindu turban. With its many depictions of magical paraphernalia and talismans, the *Catalog* was studied with awe by practitioners. In parts of the English-speaking Caribbean, people would club together to make a mail order from the *Catalog*, and it was considered a book of magic in its own right.

ALLEGED ANCIENT KNOWLEDGE!
BELIEVED LOST TO THE WORLD!!
Can Now Be YOURS!

WHO DARES TO SAY THE DAY OF MIRACLES IS PAST! WHO DARES TO DENY THAT THE POWER OF GOD IS EVERLASTING!!

What was the great power that Moses Possessed that enabled him to Practice Magic before the High Priests of Egypt? Magic so powerful that he was able to bring down upon them, the Ten dreadful Plagues, to open the Red Sea so the Hebrews could cross Safely. What were the mysteries of Solomon's Temple, the Secret rites and Mystic Proceedures that made him the Personification of WISDOM even unto our PRESENT DAY? These questions are answered in the Ten Lost Books of the Prophets compiled . . . .

FROM MUSTY SCROLLS & TABLETS BELIEVED TO HAVE BEEN FOUND IN A CAVE ON MOUNT SINAI!!

and revealed to you in all its' Mystical Splendor. The MIRACLES of Christ and how he was able to Heal the Sick, Restore Sight to the Blind, Make the Lame Walk and the Dead to Rise . . . .Be Transformed by the Renewal of Your Spiritual Gifts! Believe in the Power and it SHALL MAKE YOU WHOLE! Teaches the use of MASTER FORCE and SPIRITUAL ILLUMINATION! Open this Treasure house of KNOWLEDGE and find within it HEALTH, WEALTH, and HAPPINESS!

Send for it Today! ORDER BY No. BK115 price $3.

THE ESOTERIC TREASURE OF A LIFETIME!!
At Last!
WAS IT THIS GREAT TOWER OF STRENGTH THAT SAVED DANIEL FROM THE LION'S DEN? --OR JONAH FROM THE BELLY OF THE WHALE?
Time and again the Bible mentions various situations wherein inexplainable MYSTERIOUS FORCES have come to the rescue of People doomed to die! What were these strange Powers that saved them? Were they natural or Supernatural? Or was it some "GOD" given strength within themselves that was indomitable and could not be destroyed? Godfrey Selig who Translated the Secrets of the Psalms from the original Ancient Manuscripts claims the book.
CONTAINS PSALMS TO WARD OFF EVIL and BRING POWER & SUCCESS
He also says that whosoever shall use the correct Psalms for its' Properly given REASON shall obtain his desires in complete fulfillment! Are you TROUBLED? Do you reach for a guiding hand then learn the Secrets of the Psalms!
Don't Delay! Act Now! ORDER BY No. BK109 $2.

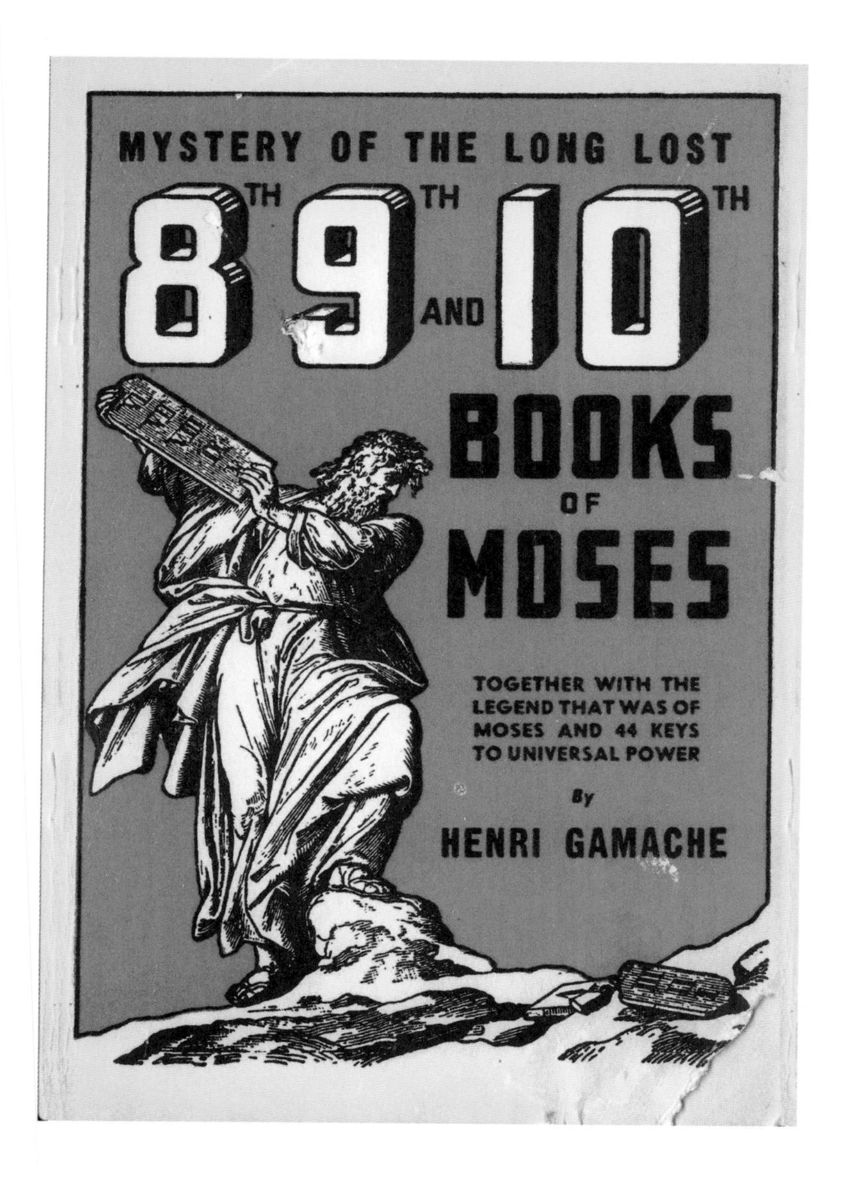

LEFT Book advertisements from the *Mystery of the Long Lost 8th, 9th and 10th Books of Moses*, New York (1948). The advertisement (top) for the *Ten Lost Books of the Prophets* by Lewis de Claremont makes it look like a sensational read, but purchasers must have been very disappointed with their $3 spend. In fact, it is simply a small-sized 32-page book that contains no practical magic at all. *The Secrets of the Psalms* (below), which was promoted as the "esoteric treasure of a life time!!," was more useful, with psalms against insanity, melancholy moods, and libel, and for keeping children alive, preventing harm from vicious dog attacks, and, of course, keeping evil spirits at bay.

ABOVE Cover of the *Mystery of the Long Lost 8th, 9th and 10th Books of Moses*, New York (1948). While the author is named as Henri Gamache, a copyright catalog from 1948 states clearly that this was a pseudonym for the author Henry Marsh. It also shows the copyright owner as Anne Fleitman, proprietor of the New York publisher Sheldon Publications. Was Fleitman the mysterious Gamache, a rare female author in the magic book trade? It shows how the venerable grimoire tradition of pseudo-authorship remained alive and well in the 20th century.

ABOVE The cover of *The 7 Keys to Power* certainly attracts the eye, with its striking use of red-and-black print on yellow paper. It includes an engraved portrait of the pseudonymous Lewis de Claremont, dressed in a turban like a great Indian adept—just as De Laurence often portrayed himself. Written in a narrative first-person style, *The 7 Keys* proves to be a familiar mix of numerology, dream interpretation, thought power, and prayers. It claimed that once mastered, "you will be able to cast out devils, put them into people, overcome enemies, uncross people, cross people and make anyone do your bidding without knowing it."

RIGHT The *Ancient's Book of Magic*, published by Dorene in 1940, was one of the more polished of the de Claremont oeuvre, although the image print quality is terrible. A cover price of $5 provided 137 pages of instructions on the invocation of spirits and a "list of angels, spirits, demons, goblins as compiled by Lewis de Clarmont [sic]." There are also blurry images of the spirits taken from Barrett's *The Magus* (see pages 124–125), or most likely de Laurence's reproductions of them from the *Great Book of Magical Art* (see page 194).

The ANCIENT'S

BOOK OF MAGIC

PRICE $5.

CONTAINING

SECRET RECORDS OF THE PROCEDURE & PRACTICE OF THE Ancient Masters & Adepts

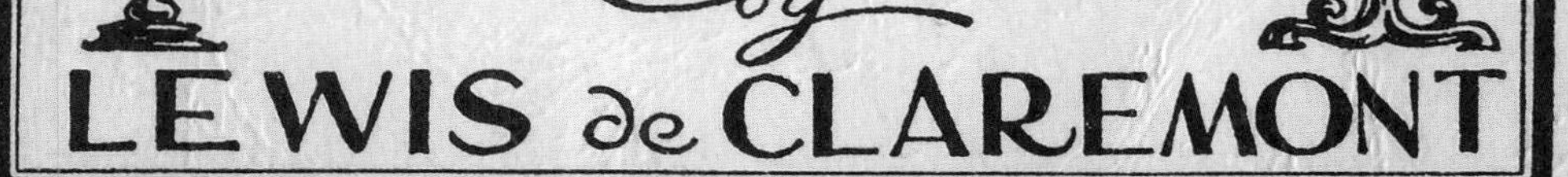

by

LEWIS de CLAREMONT

William Lauron De Laurence

Few people today have heard of William Lauron De Laurence (1868–1936), and yet he was a magical legend in his own lifetime for some in America, the Caribbean, and parts of West Africa. As was stated proudly in one of his catalogs, "His works are everywhere and his name stands high on the roster of independent and intrepid thinkers of all time."

De Laurence was born in Cleveland, Ohio, to a Pennsylvania Dutch mother and a French-Canadian father. As a teenager, he worked on the railways as a flagman, and then turned to door-to-door salesmanship selling books on hypnotism and the like. He later set up his own short-lived school of hypnotism before creating a publishing company. He knew from experience that there was a public appetite for writings on the occult, and he had his first real hit with a version of the mystical book of talismans, the *Sixth and Seventh Books of Moses*. He later recalled, "I bought the book because there were many people selling it at the time for $5.00 a copy and I supposed it would be a good seller."

More re-editions and plagiarized magic works followed, including *The Great Book of Magical Art, Hindu Magic and East Indian Occultism*, which was essentially a copy of Francis Barrett's *The Magus*. Then, De Laurence spotted the potential of the recent booming mail-order market, and he began to use domestic and international newspapers to advertise his postal wares. An annual *De Laurence Catalog* appeared, offering a huge range of magical books, talismans, and equipment, and quickly became established as an essential resource for the

RIGHT Portrait of William Lauron De Laurence (c. 1910). De Laurence sometimes wore a turban in portraits, claiming to have been adopted as a Hindu *swami* or religious teacher. With the illustrative border on this portrait, however, he was tapping into the iconography of mystical ancient Egyptian knowledge, which was a growing aspect of African American empowerment.

LEFT Cover of *De Laurence's Catalog*, Chicago (1940). The annual *Catalog*, which De Laurence advertised in Caribbean and Nigerian newspapers, was key to his success in developing a thriving overseas mail order business. In 1908, the *Catalog* was a slim 48 pages but by the 1920s, it had grown to 60 pages, and by 1940, it was a whopping 576 pages.

budding magician. Young men even traveled from West Africa to the United States to learn from De Laurence, and work at his premises in Chicago.

By 1919, De Laurence estimated that he had around $100,000 in plates, stock, and merchandise, and claimed he was the greatest publisher of occult and magical works the world had ever known. His plans to open a store were put on hold when he was prosecuted by the American postal service for "conducting a scheme for obtaining money through the mail by means of false pretences, representations, and promises." Yet, his business continued to prosper nevertheless, and by the time he died in 1936 even his catalog had become incorporated into the practices of some West African and Caribbean religious and magical movements. This expansion into the Caribbean stalled in 1940, however, when the British colonial authorities in Jamaica prohibited all works by De Laurence, and the ban remained in place after independence.

The Master Book of Candle Burning

OR

How to
BURN CANDLES
For Every Purpose

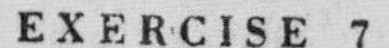

EXERCISE 7

TO GAIN POWER OVER OTHERS

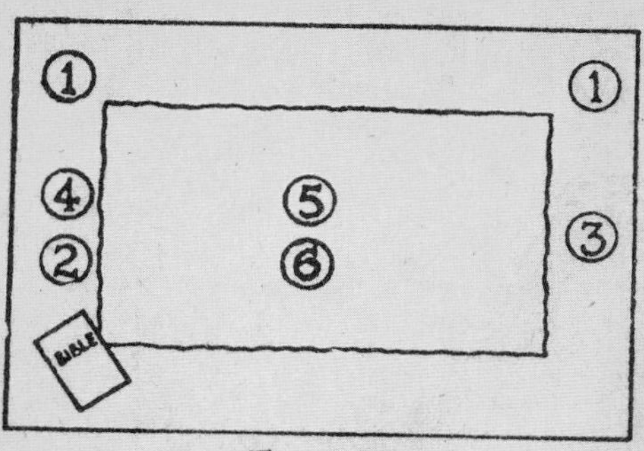

Fig. 19

Position 1-1 *Black Crucifix Altar Candles dressed with XX DOUBLE CROSS brand Oil.*
Position 2 *Astral Candle of petitioner dressed with Zodiac brand Oil.*
Position 3 *Astral Candle of person over whom power is sought dressed with Zodiac brand Oil.*
Position 4 *White Candle (or Flaming Power brand Candle) dressed with Power brand or Domination brand Oil.*
Position 5 *Purple Candle (or Conquering Glory brand Candle) dressed with Conquering Glory brand Oil.*
Position 6 *Orange Candle (or Crown of Success brand Candle) dressed with Crown of Success brand Oil.*
Movement: Move candles 5 and 6 toward candle 3 two inches each day.
Suggested Bible Reading: Psalm No. 130
Devotions: 1 hour nightly for six days or longer if desired.

84

LEFT AND ABOVE *The Master Book of Candle Burning*, New York (1942). As stated in the first chapter: "It makes little difference what your religion may be—candle burning brings consolation and solace. Candle Burning perhaps better illustrates the UNIVERSALITY of MAN than any other thing." The theory behind candle burning was based on some basic reading of the folklore of ancient fire rituals. The author, Henri Gamache (see page 197), even provides a bibliography, which includes the anthropologist James Frazer's famed book on comparative belief systems, *The Golden Bough* (1890). As to practice, the image above is one of twenty that show how to position candles of different types (some purchasable magical brands) in a room. As well as for gaining power over others, the candle magic diagrams were for conquering fear, bringing good luck, arousing discord, and protecting against evil influences.

ABOVE AND RIGHT *Leyendas de la magia del incienso hierbas y aceite* (*Legends of the magic of incense herbs and oil*), New York (1938). This manual was apparently inspired by de Claremont's spirit guide, "Appollonius of Tayaneus"—otherwise known as the Greek philosopher Apollonius of Tyana (c. 3 BCE–c. 97 CE)—who visited India in search of knowledge, according to ancient legend. The manual contains the usual de Claremont miscellany of poor cut-and-paste images, thinking aloud, and practical instructions. There are a series of sketchy images of how to insert talismans into one's mattress, clothes closet, and shoes. The curious diagram (above) is associated with instructions on anointing oneself with special oil to call up and attract the spirits, although it was advised to wash the oil off in a bath straight afterward.

LEYENDAS
DE LA
MAGIA
DEL
INCIENSO
HIERBAS Y ACEITE
por
PRECIO: $2.00
LEWIS de CLAREMONT

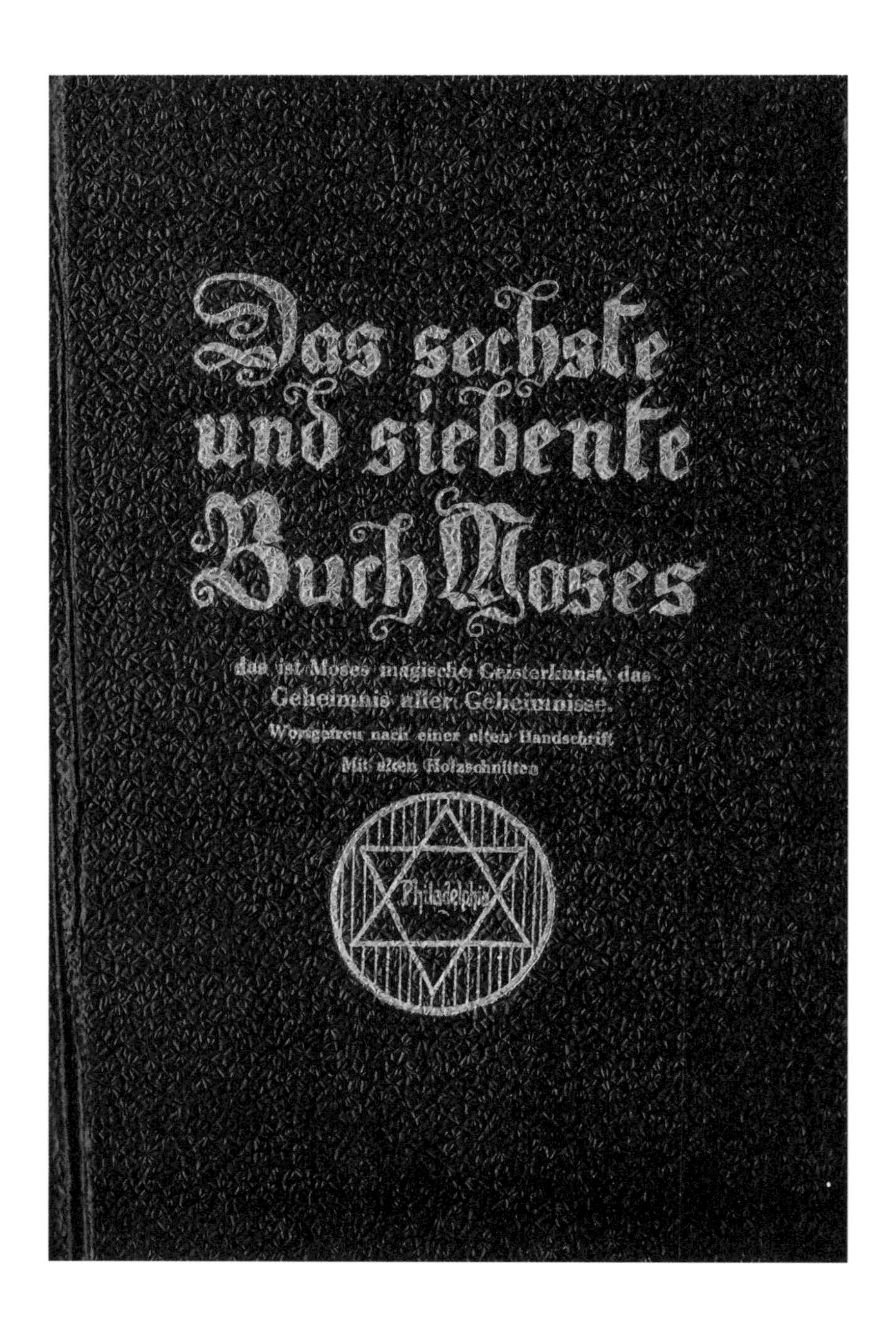
Das sechste
und siebente
Buch Moses
das ist Moses magische Geisterkunst, das
Geheimnis aller Geheimnisse.
Wortgetreu nach einer alten Handschrift
Mit alten Holzschnitten
Philadelphia

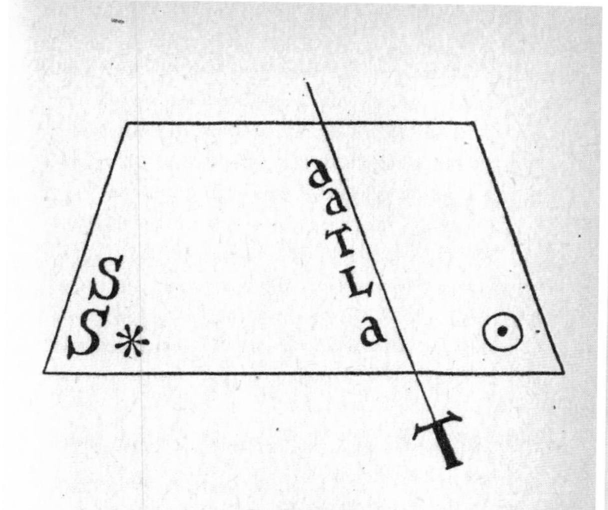

LEFT AND ABOVE *Das Sechste und Siebente Buch Moses* (*Sixth and Seventh Books of Moses*), Braunschweig, Germany (1950). Although the title page suggests it was published in Philadelphia, this 20th-century German version of the Moses Books bears little relation to the *Sixth and Seventh Books of Moses* that circulated in America. There are few seals and talismans, and little in the way of pseudo-Hebraic imagery. The purported Jewish origins of its content is toned down, perhaps in response to the Holocaust. There are a couple of seals of Lucifer, one of which is copied from the *Dragon Rouge* (see page 112), while the amulet shown (above left) is a "Moses blessing" for protection against spirits.

श्री रुद्र यंत्रम्

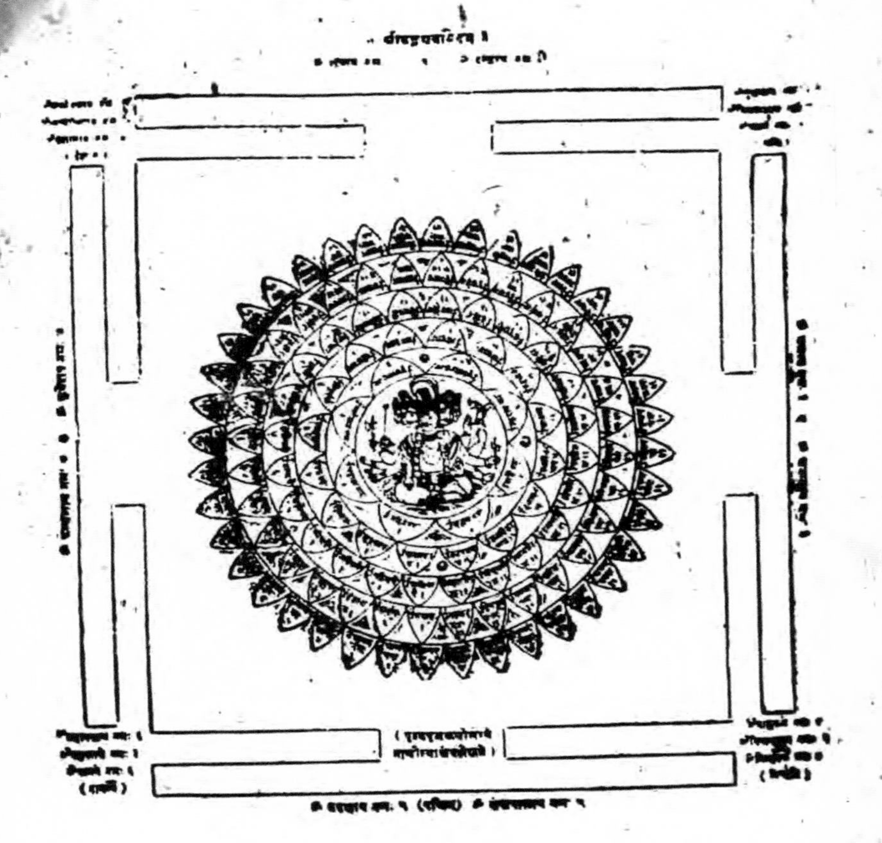

ॐ नमः शिवाय, ॐ नमः शिवाय

इसे भोजपत्र में लिखकर ताबीज बनाकर बाँधने से शिवजी की विशेष कृपा तथा धन-धान्य की वृद्धि होती है ।

LEFT AND ABOVE Pages from the *Brihat Indrajala* (*The Big Art of Magic*), which was published in Hindi (see page 179). The word *indrajala* derives from the ancient Indian notion that a magician or healer makes a net (*jala*) over the sensory organs (*indriyas*), the tools of knowledge and wisdom, and thereby affects the cure of both mind and body. This book runs to several hundred pages and contains a range of incantations, talismans, magic squares, mantras, and yantras. The image (left) is inspired by the medieval *dhāranīs* (Buddhist mantras) with the eight-armed goddess Mahāpratisarā (see page 63). There are numerous illustrations of the Hindu gods. The image (above) represents Lord Kala Bhairav, the God of Time, and also an expression of Shiva, who has often been depicted riding on a dog. He is a punisher of sin and is worshipped for good fortune and protection against black magic.

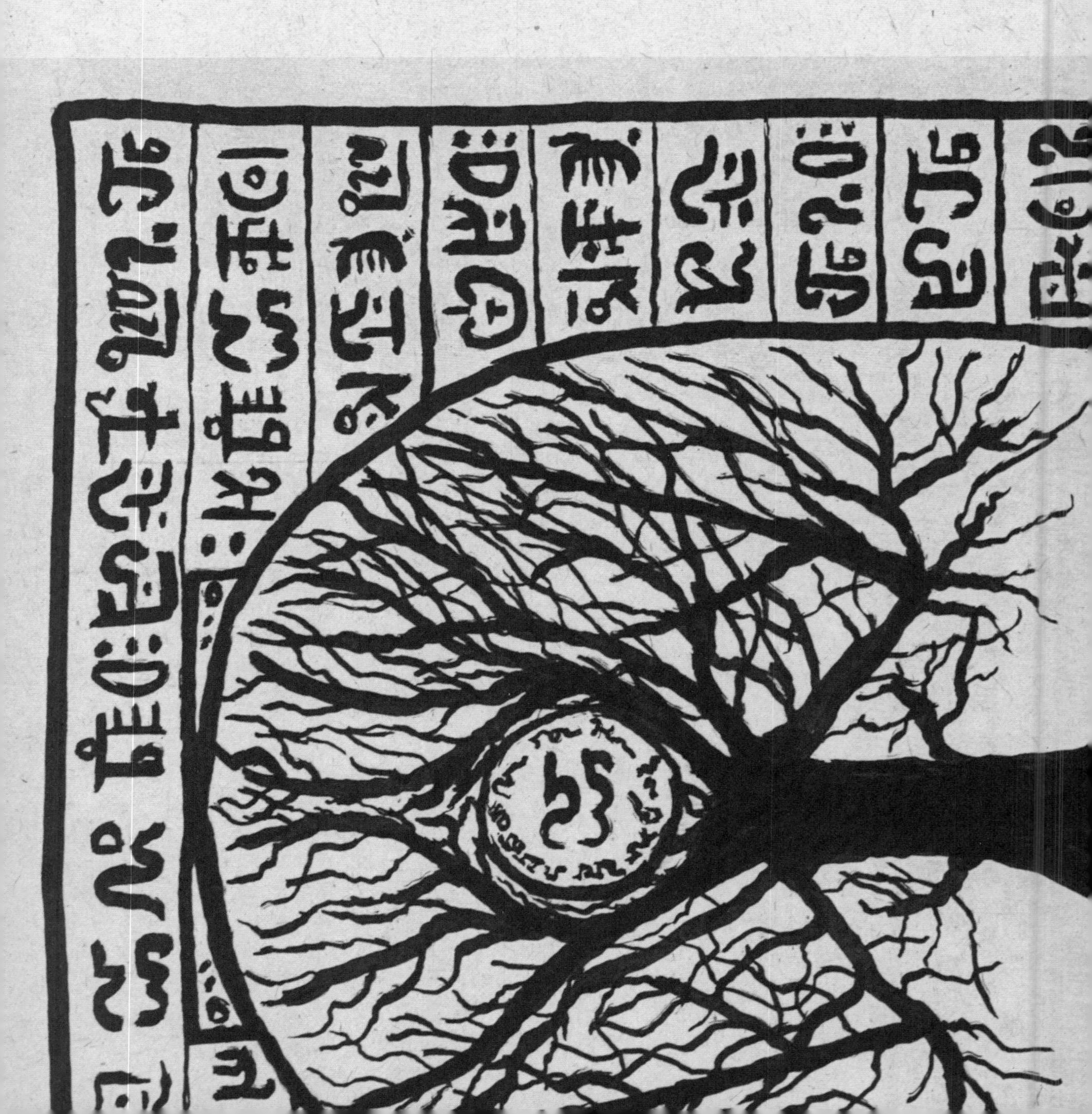

CHAPTER SIX

The Contemporary Grimoire

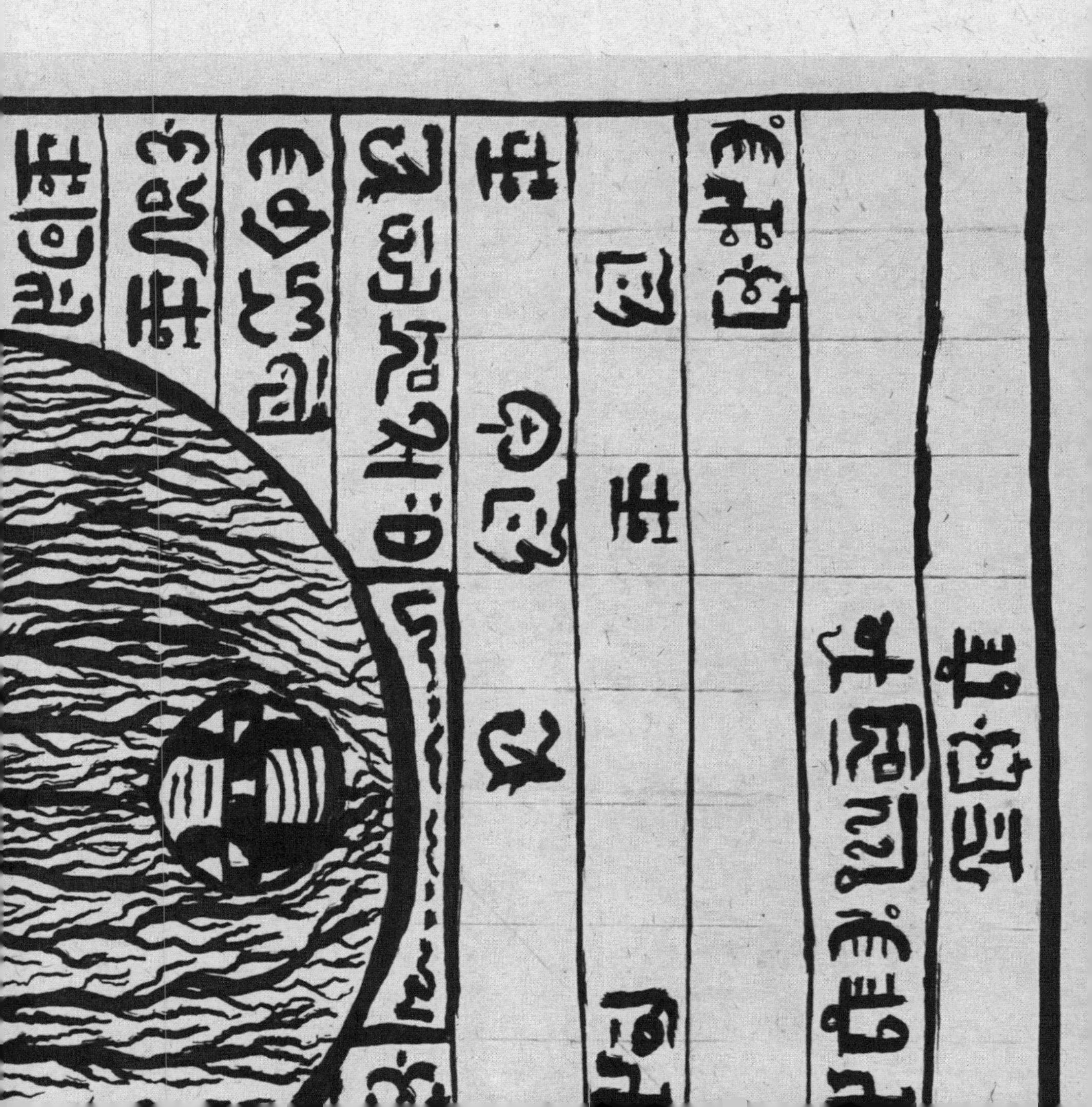

A s cheap printed magic books swirled around the world, in late 19th-century Britain a small group of middle-class men and women created a new secret, magic movement in the form of the Hermetic Order of the Golden Dawn, dedicated to Renaissance forms of mystical alchemy and kabbalah, as well as astrology, tarot, and geomancy. Their understanding was garnered from manuscripts and early-modern printed texts, and supplemented by the magical knowledge recently unlocked through archaeological discoveries from the ancient Egyptian and Hellenistic worlds. The founders of the Golden Dawn movement were Freemasons, and the hierarchical structure of the Order and its initiations were based on masonic organizational principles. While never numbering more than a few hundred, the Order attracted influential figures in the artistic world, such as the Irish poet W.B. Yeats (1865–1939). Furthermore, the magic books published by some of its members would go on to inspire the imaginations of many artists, musicians, writers, and film makers across the Western world right up to the present day.

Samuel Liddell Mathers (1854–1918), one of the masonic founders of the Golden Dawn, published the first English print edition of the *Key* or *Clavicule of Solomone*. He pieced it together from several manuscripts in different languages, mostly 17th-century examples held in the British Library, which, in turn, borrowed from late medieval grimoires of spirit conjuration. First published in 1889, it has been through dozens of editions and is still in print today. In typically outrageous fashion, De Laurence also published an American edition in 1914 without permission, ensuring that the book reached a much wider and culturally diverse international readership through his catalogs than Mathers intended. Mathers produced another translation the following decade, this time of *The Book of the Sacred Magic of Abramelin the Mage*, a French manuscript from around 1700 kept in the Bibliothèque de l'Arsenal in Paris, and supposedly the work of a medieval Egyptian adept. While Mathers practiced ritual magic with a passion, another leading member of the Golden Dawn, Arthur Edward Waite (1857–1942), was far more interested in Christian mysticism. Mathers and Waite were the chalk and cheese of the Golden Dawn, and Waite dismissed Mather's publications as arid material for occult insight.

LEFT A portrait by Moina Mathers (1865–1928) of her husband Samuel Liddell MacGregor Mathers (1854–1918) in ceremonial regalia, oil on paper (c.1895). Moina Mathers, whose birth name was Mina Bergson, was a trained artist and fellow member of the Golden Dawn. The couple performed together in ritual costume. The Egyptian inspiration for the headdress is clear, with the pentacle or Seal of Solomon also prominent.

RIGHT A painting by Aleister Crowley, entitled *The Hierophant*, oil on board (1921). The term "Hierophant" comes from the ancient Greek, meaning the priest or priestess who reveals the sacred mysteries to their followers. There is a degree of self-portrait in this image, as shown by the diabolic 666 prominent on the chest. Crowley liked to style himself as the "Beast 666," and revelled in being labeled the "Wickedest Man in the World" by the press.

LEFT Some of the ritual manuscripts used by Doreen Valiente, including Gerald Gardner's *Book of Shadows*. Although rather publicity shy compared to Gardner, Valiente wrote three influential books during the 1970s on her approach to Wiccan magic and ritual, including *An ABC of Witchcraft* (1973). These were produced by the well-known, mainstream publisher Robert Hale.

Waite's contribution to the modern imagination lies in the publication of the Rider Waite tarot deck, illustrated by fellow Golden Dawn member Pamela Colman Smith (1878–1951), and which is the most widely used tarot pack today. He was also famed for his compendious collection of conjurations, the *Book of Black Magic and of Pacts* (1898). Subsequently published in a cheaper edition with the title, *The Book of Ceremonial Magic*, this was the largest compilation of conjuration rituals ever published in English, with Waite drawing upon the likes of the pseudo-Agrippa *Fourth Book of Occult Philosophy* and French cheap print grimoires such as the *Grand Grimoire*, *Dragon Rouge*, and *Grimoire du Pape Honorius*. For Waite, such books were magical trash and, like Reginald Scot three centuries earlier (see page 93), his purpose for printing their contents was to show their "absurd" and "iniquitous" nature. Of course, as a prolific author of books on mysticism and alchemy, Waite was also looking to make a bit of money.

The notorious Aleister Crowley (1875–1947) was another important figure publishing old magic during this early 20th-century Magical Revival. As a youthful member of the Golden Dawn, Crowley was mentored by Mathers but both were in the possession of big egos, and they eventually fell out. Crowley went on to cofound his own ceremonial magic group the A∴A∴, also known as the Argenteum Astrum. As an act of revenge after their argument, Crowley published one of Mather's transcribed manuscripts without permission—this time a 17-century grimoire held in the British Library that included a list of spirits, parts of which were evidently copied from Reginald Scot's *Discoverie* (see pages 93 and 118–119). Although hardly a bestseller, *The Book of the Goetia of Solomon the King* (1904) injected the word "goetia"—meaning the summoning of spirits—into the creative consciousness.

One of those inspired by all these works was a former colonial civil servant named Gerald Gardner (1884–1964). He claimed to have been initiated into a secret, pagan English witch cult that had survived centuries of persecution. According to Gardner's story, he was given the ancient rituals and knowledge of the coven in the form of an old manuscript called the *Book of Shadows*. This became the founding text of his new pagan religion, known as Wicca. It was the sort of discovery story seen many times before in the history of grimoires. The physical book that he subsequently showed to his coven members was proven to have borrowed content from both Mathers' *Key of Solomon* and Crowley's rituals. The *Book of Shadows* that came to be used widely by Wiccans was a re-write by the High Priestess of Gardner's coven, Doreen Valiente (1922–1999). She set about cutting out what she called the "Crowleyanity," and added her own distinctive creative content. Several decades on from the founding of Wicca, contemporary Paganism would develop into diverse strands, and the idea of the solitary practitioner grew stronger and stronger. Rather than be tied to coven hierarchies and single texts, modern witches began to create their own *Books of Shadows*, inventing their own rituals and spells from published books of magic, both old and new.

THE MAGIC REVIVAL AND THE CREATIVE INDUSTRIES

The wider cultural influence of the Golden Dawn and Wicca starts to become noticeable during the counterculture movement of the 1960s and early 1970s. It can be heard (and seen) in psychedelic and heavy metal music, for example—particularly Crowley's influence, due to his use of sex and drugs in magic ritual. Satanism also became a theatrical rock-and-roll movement in Anton Szandor LaVey's American Church of Satan, an organization that rejected Christian hypocrisy and, despite its name, espoused worship of the self rather than the Devil. As a result of this magical influence, strange creative mashups began to occur, and there is no better example than the light-hearted comedy classic film *Bedknobs and Broomsticks* (1971).

LEFT *Book of Shadows Spell Template*, a commercial digital download to produce printable blank spell pages. The digital age takes the act of writing magic in a new direction from the days of papyrus and parchment. Is it still magic if a desktop printer does part of the job for you? Why not. It is simply the latest technology to facilitate the recording of spells.

ABOVE The *Book of Shadows* was the most recognizable prop in the television series *Charmed*, appearing in nearly every episode. It was essential to the magical power of the three Halliwell sisters, and the first episode of the series was called "Something Wicca This Way Comes" (1998).

Set in a British coastal village in 1940 under the threat of Nazi invasion, Angela Lansbury plays spinster Eglantine Price, who unbeknown to her neighbors is taking mail-order lessons in magic in order to help protect her community. These lessons are based on an old grimoire called the *Book of Astaroth*, named after a great sorcerer from the Isle of Naboombu. Price finds out that her teacher, Professor Emelius Browne, is a charlatan who only possesses half the book. So, she and her child pupils set off on a magical adventure to find a key spell that is inscribed on the Star of Astaroth talisman. The film deviated significantly from the source material of Mary Norton's books, published in the 1940s, in which Astaroth does not appear and Emelius is a time-traveling, 17th-century sorcerer. However, it's a good example of the way in which magical history began to influence the creative industries.

Astaroth was a name to conjure with and to excite the imagination. Astaroth was listed as a great demon in such early modern works as Weyer's *Pseudomonarchia Daemonum* (see page 93). He was also mentioned by pseudo-author Dr. Faust (see page 102), and he was referenced in the *Book of Abramelin* and *The Book of the Goetia*. It is likely that the scriptwriter for *Bedknobs*, the veteran Disney writer–producer Bill Walsh, got the name from Waite's *Book of Ceremonial Magic*, in which there are numerous references to the demon, including instructions for "The Conjuration of Astaroth." An American edition of *The Book of Ceremonial Magic* had been published in New York in 1961 by University Books, which had also published Waite's popular *The Pictorial Key to the Tarot*, two years earlier. University Books had been set up in the mid-1950s, along with the associated Mystic Arts Book Society, by the intriguing Felix Morrow

(1906–1988) to promote new and old works on mysticism and occultism. Born into a Hasidic family, Morrow was a former Communist, who grew disillusioned by Stalin's Great Terror.

In the multimedia world of the late 20th and 21st centuries, Astaroth has been referenced in a number of horror films, and was also the name of an evil demon king in the successful *Ghosts 'n Goblins* computer game series created in the mid-1980s by Japanese company Capcom. The word "goetia" continues to resonate, too. It was the title of a computer game, released in 2016 by French studio Sushee, in which the player takes on the role of the ghost of a young girl. "Stella Goetia" is also the name of a powerful demon in the American adult animated comedy *Helluva Boss* (2020), which concerns an assassination company based in Hell. The title *Book of Shadows* proved catchy and came to be a household name through references on television and cinema in the age of the teen witch phenomenon of the 1990s. It was mentioned by the spell-casting schoolgirls in *The Craft* (1996), and it was the name of the big book of magic regularly used by the three modern witch sisters in the hit television series *Charmed* (1998–2006).

LEFT Promotional image for *Le Grimoire d'Arkandias* (2014), a French feature film based on the novel of the same name by Eric Boisset (1996), which was published a year before the first *Harry Potter* novel appeared. One of the main child characters, Théophile, as part of a trio of questing youngsters, comes across a book in his local library entitled *Practical Lessons in Red Magic*, from which falls a note reading, "How to Become Invisible, by A. Arkandias." And so, the adventures begin. The spell for the invisible ring includes the blood of a black hen—a reference to the infamous ritual of the *Poule Noir* (see page 112).

HORROR AND FANTASY

While books of magic have often been portrayed as a source of female empowerment in the modern witch genre, the grimoire has also provided artistic inspiration in the realms of horror. In the Japanese film *Corpse Party Book of Shadows* (2016), the search for the eponymous forbidden book by a group of students leads to gory death scenes, and when the book is finally opened to prevent further murders, the results are not exactly what was hoped for. In *Warlock* (1989) the sought-after book of dangerous black magic is called the *Grand Grimoire*. The time-traveling warlock is a satanic figure, who roams the past and present searching for the three separated sections of the *Grand Grimoire* in order to destroy the world. Of all the horror grimoires the *Necronomicon* is the most iconic, featuring in the *Evil Dead* films but also in video games and comic books, and it was parodied by Terry Pratchett, author of the *Discworld* series, with his *Necrotelecomnicon* or "Telephone Book for the Dead." The *Necronomicon* was the creation of the seminal horror writer H.P. Lovecraft (1890–1937), who said the name came to him in a dream. He then set about creating its fictional author, the 8th-century, Arabic-sounding and half-mad Abdul Alhazred. In Alhazred's time, his book was known as the *Kitab al-Azif*, but just as the *Ghāyat al-Ḥakīm* became known as the *Picatrix* in the West (see page 55), so, too,

BELOW The *Monster Book of Monsters*, from Warner Bros. Studio Tour London: The Making of Harry Potter. It is an example of one of numerous physical creations imagining the look of the *Monster Book of Monsters* (including in LEGO), which is a text assigned by Professor Hagrid for his trainee wizards in *Harry Potter and the Prisoner of Azkaban*. But the book has sentience, it is a monster in its own right, refuses to open, and bites the fingers of those that try without performing the necessary ritual petting.

the *Kitab al-Azif* became the *Necronomicon*. After inventing the *Necronomicon*, Lovecraft read Waite's *Book of Black Magic*, which gave him ideas about its possible contents. Lovecraft liked the idea of other fantasy-horror authors referencing and borrowing his inventions, as it helped enhance the unsettling sense of realism in his imagined world.

This idea of collective creative development would later be fully borne out in the multimedia universe of modern fantasy, where role-playing, film, television, and computer games provide a rich immersive experience, and the original stories are built upon and expanded by fans and creatives alike. Magic books act as bridges between old and new material, and help take stories forward in new directions. In the world of *Conan the Barbarian*, originally created in the 1930s by the pulp fiction writer Robert E. Howard (1906–1936), there is the iron-bound *Book of Skelos* that contains the ancient secrets of the wizards of Atlantis. Dating to the Hyborian Age on Conan's Earth (between 14,000 to 9500 BCE), it was used to summon extra-terrestrial demons and to resurrect the dead. This thread was followed up in an episode of the animated *Conan: The Adventurer* series (1993), when a quest for a "Second Book of Skelos" takes place, after the first-known example has been destroyed. Another successful multi-platform fantasy world with game products, artwork books, and novels, called *Dragonlance*, plays heavily on the power of wizards and their spell manuscripts. Books of magic become power-ups, points, and weapons, as well as plot devices in the gaming world.

The library of magic, as distinct from the individual grimoire, is a motif that cuts across modern genres of fantasy, representing not only the repository of ancient wisdom but acting as a physical portal between the real world and the supernatural realm. It is a place where monsters may lurk, and where magical things are not only learned but also occur. The library is central to the magical interplay in the television series *Buffy the Vampire Slayer*, for instance, and the Hogwarts Library is often included in the action in the *Harry Potter* series of books. In the *Dragonlance* fantasy world, there is reference to a collection of ancient magic in the Great Library of Palanthas and also an archive of books on the history of magic and magical manuals in the Tower of Wayreth. In our contemporary world, where the internet provides the largest repository of magical works ever created, the idea of a physical library of magic, whether real or imagined, with its secrets, atmosphere, and possible revelations, continues to excite. The library of magic can be more than the sum of its books.

In one of numerous clever footnotes in Susanna Clarke's novel *Jonathan Strange & Mr Norrell* (2004), which concerns learned magicians, their books, and their libraries in Napoleonic-era England, the narrator comments, "a book of magic should be written by a practicing magician, rather than a theoretical magician or a historian of magic. What could be more reasonable? And yet already we are in difficulties." As this book has shown, these difficulties readily apply to magic books throughout history—and no more so than today. Books of magic have continually been invented and imagined, as well as provided a record of occult empiricism and accumulated wisdom. They have become aesthetic objects in their own right, with contemporary notions of what magic books should look like and what they mean shaped by modern reinvention, conceptions of the past, and dreams of other worlds. Artists and historians can also be magicians in their own way.

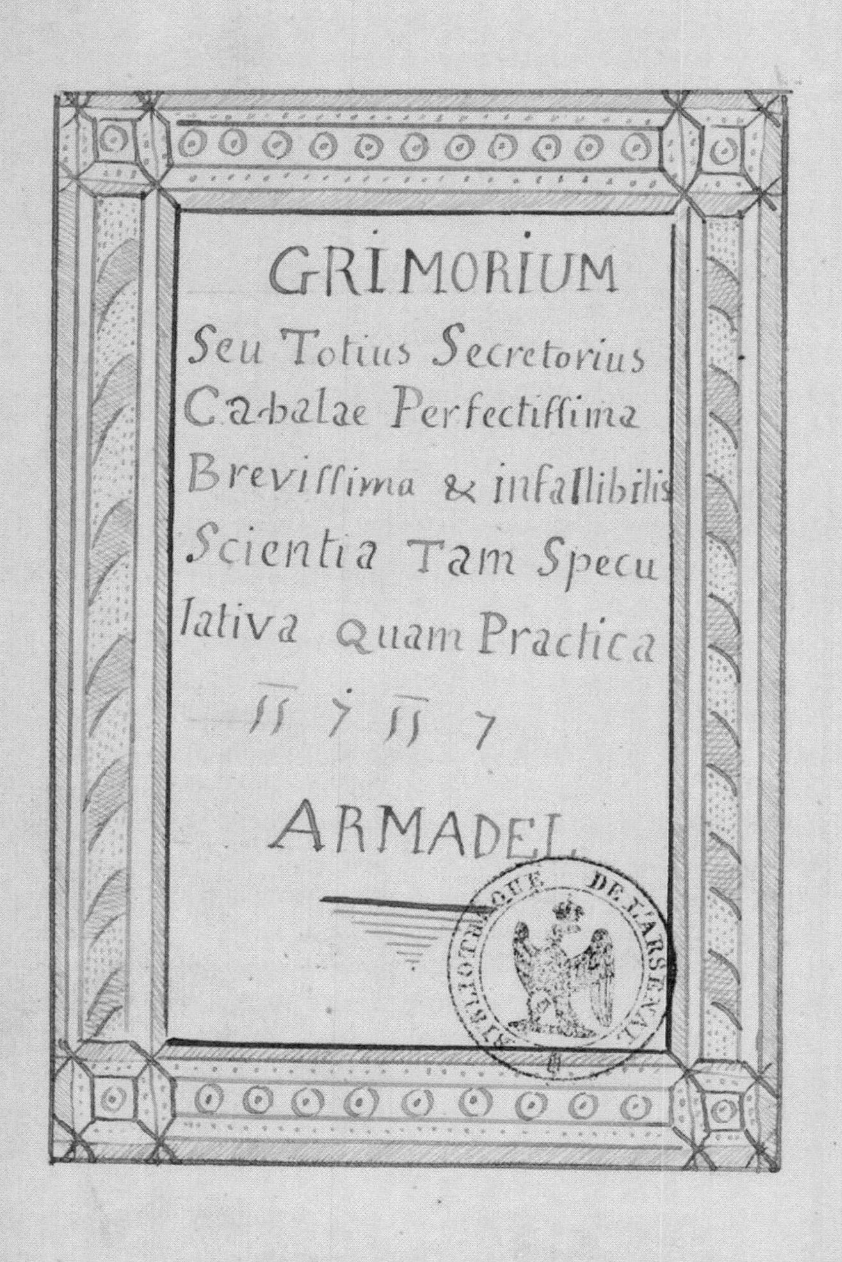

GRIMORIUM

Seu Totius Secretorius
Cabalae Perfectissima
Brevissima & infallibilis
Scientia Tam Specu
lativa quam Practica

ה י ה ו

ARMADEL

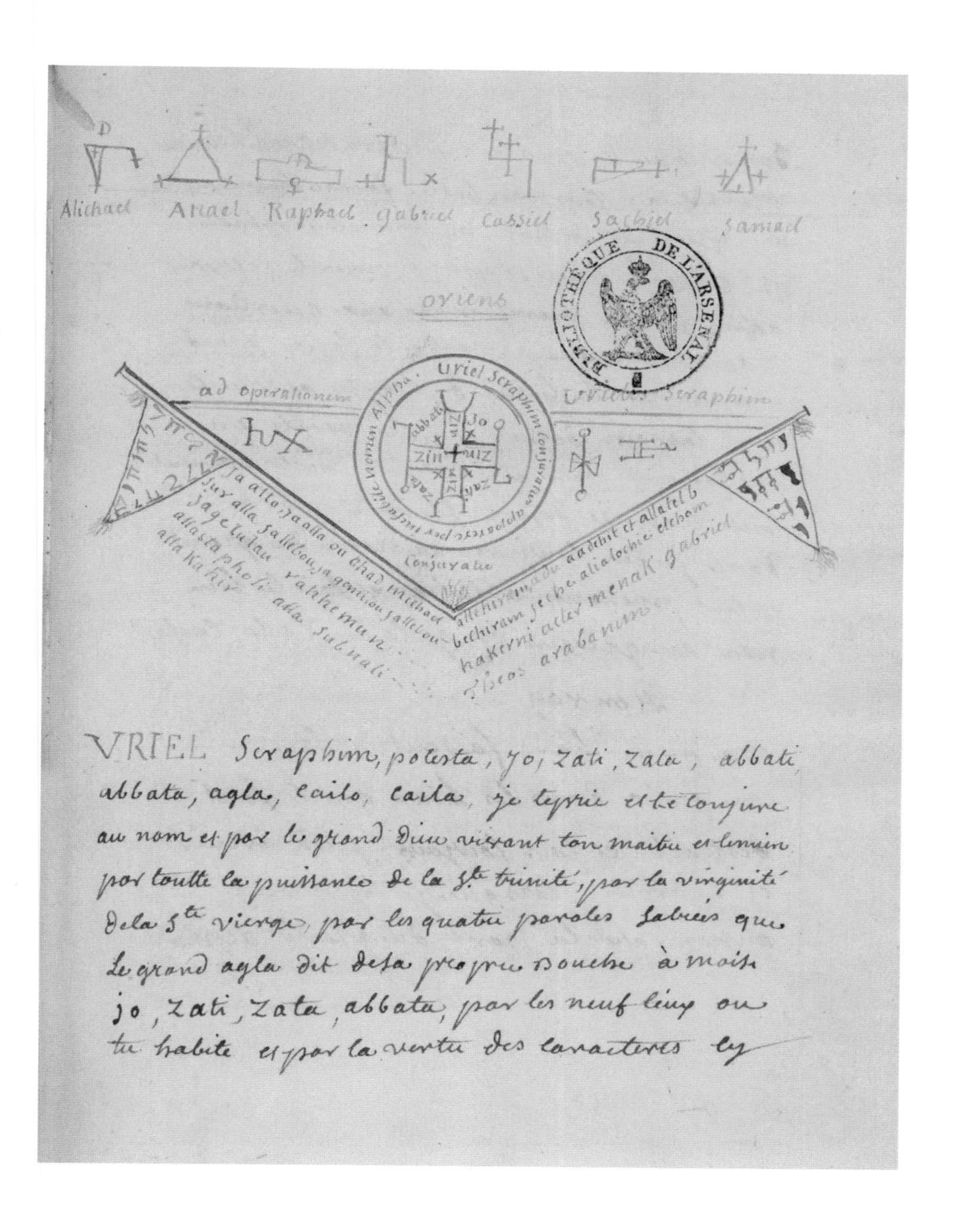

VRIEL Seraphim, potesta, Jo, Zati, Zala, abbati
abbata, agla, caïlo, caila, je te prie et te conjure
au nom et par le grand Dieu vivant ton maître et le mien
par toutte la puissance de la Ste trinité, par la virginité
de la Ste vierge, par les quatre paroles sacrées que
Le grand agla dit de sa propre bouche à moïse
jo, Zati, Zata, abbata, par les neuf lieux ou
tu habite et par la vertu des caracteres cy

LEFT AND ABOVE Title page and illustration from the *Grimorium, seu totius secretorius cabalae perfectissima, brevissima et infallibilis scientia, tam speculativa, quam practica Armadel* (*Grimoire, or the most perfect, concise and infallible science of the entire secret Kaballah, both theoretical and practical Armadel*), written in French and Latin on paper (18th century). MacGregor Mathers translated this into English during his time researching the collection of grimoires in the French Bibliothèque de l'Arsenal, but it was only published in 1980. The Bibliothèque de l'Arsenal has recently made a digitized copy available to all. Once closely guarded, secret manuscripts, which passed from hand to hand, are disseminated freely now on the internet as part of a new democratic digital magical tradition.

ABOVE An unattributed, and artificially aged, modern grimoire manuscript. The imagery is largely decorative, and has little magical coherence. The text shown here is actually a poem by Aleister Crowley entitled "The Poet":

Bury me in a nameless grave!
I came from God the world to save.
I brought them wisdom from above:
Worship, and liberty, and love.
They slew me for I did disparage
Therefore Religion, Law and Marriage.
So be my grave without a name
That earth may swallow up my shame!

ABOVE One of the earliest surviving "Book of Shadows," created by Ralph Harvey between 1955 and 1957. Harvey (1928–2020), was an escapologist and stuntman in his professional life, and was an early adherent of Wicca. He established his own Wiccan group, the Order of Artemis, in 1959, and was the author of several books later in life, including *The Last Bastion: The Suppression and Re-Emergence of Witchcraft, the Old Religion* (2004). He gave the eulogy at the funeral of Doreen Valiente in 1999.

ABOVE *Bird Men of Burnley* by Leonora Carrington, oil on canvas (1970). This absorbing painting appears to be inspired by the old alchemical emblem books, and the imagery of the conjurations performed by witches and magicians in early modern art. Born into a wealthy English family, Carrington (1917–2011), who lived most of her adult life in Mexico City, was an early British exponent of Surrealist art who focused on female sexuality, as well as magical and alchemical symbolism in her paintings and prose. She read books on the history of esotericism, folklore, Eastern mysticism, the psychology of the occult, and the works of Gerald Gardner on Wicca. One of Pamela Colman Smith's illustrations for the Rider-Waite tarot pack inspired her portrait of her one-time lover, the German Surrealist artist Max Ernst.

ABOVE Posthumously published in 1956, *Moon Magic* was one of several works of occult fiction written by the British ritual magician Dion Fortune (1890–1946). Born Violet Mary Firth, she grew up in a wealthy Christian family and for much of her life Fortune's views on, and practice of, magic were primarily guided by Christian mysticism, although she would later come to embrace aspects of Pagan worship. In the 1920s, she was a member of Moina Mathers' spin-off Golden Dawn group, the Alpha et Omega, but the pair fell out with each other. By this time, Fortune had already created her own organization, the Fraternity of the Inner Light. The plot of *Moon Magic* reflected her later Pagan interests, and was an unfinished sequel to her novel *The Sea Priestess* (1938), which included practical accounts of "ancient" rituals.

Sigil Generation

The English artist and occultist Austin Osman Spare (1886–1956), who drew upon Buddhism and Taoism, and who also experimented with the application of automatic writing to unlock the unconscious expression of art, attempted to re-think sigils for a new artistic age. Spare's "sigilization" theory, published privately in his *Book of Pleasure* (1913), held that "sigils are the art of believing; my invention for making belief organic, ergo, true belief." As a technique, it involved writing a short sentence expressing one's desire, which was then reformulated creatively to form a single glyph that would work on a subconscious level to calm conscious emotional clashes. As someone immersed in the Magical Revival, Spare was clearly artistically inspired by the sigils he saw in such venerable works as the *Key of Solomon* and Agrippa's *Three Books of Occult Philosophy*, but for him the creation of sigils was not about ritual communication with the spirit realm but rather about tapping the personal subconscious.

Spare was involved in Aleister Crowley's Argenteum Astrum for several years in the early 20th century but disliked ceremonial magic and tired of Crowley's ways. His theory of sigilization would gain much more influence after his death in the "chaos magic" movement that emerged in Britain in the 1970s. The premise of chaos magic, or magick, is the rejection of rigid, historically-defined, mechanical rituals, and the dismissal of the notion that these lead to ultimate truth. Instead, practitioners focus on the essential techniques of *doing* magic through symbol systems. This makes it a versatile tradition, with its various branches finding inspiration in an eclectic way, borrowing from the I Ching, Kabbalah, the Enochian language, and mantras, as well as personal sigil creation. Each practitioner finds his or her own path to achieve their personal development and psychic goals.

It is Spare, via chaos magic, who inspired the designers of Sigil Engine (*sigilengine.com*), a digital arts project devised "to make the process of creating sigils authentic and magickal, through a carefully considered artistic process. Art is, of course, magick and this project is a participatory piece of art." It went live in 2020 and within months the site had generated over two hundred

LEFT The sigil for "wishing for peace" generated by the Sigil Engine (www.sigilengine.com). The designers explain: "The Sigil Engine was born of many discussions about the potency of Technomancy as a form of magick. We decided the only way to find out would be to build a Sigil generator."

thousand sigils. They are not the only ones offering such "technomancy" on the internet. Another free sigil generator site, *psychicscience.org*, is similarly inspired in its origin and aims, stating its "system and algorithms" are "based on traditional magical methods."

ABOVE The sigils for "magical protection" (top) and "conjuring spirits" (bottom) created by the Sigil Magic Generator (psychicscience.org). The website advises: "Simply type in your intention or other key idea (such as a concept or personal name), then watch as your personalized sigil is drawn. You should aim to make your intention as specific as possible."

ABOVE "The Alchemist's Blue Grimoire". Created by Québec artist and bookbinder Étienne Milette, 2018. In recent years there has been a growing market for props and art objects that represent aged and distressed books of magic. Some elaborate prop grimoires are created from foam, glue, and paper, but Milette produces hand-crafted, weathered leather covers like those of the Blue Grimoire, and between them are stitched leaves of artificially aged paper using tea-staining. "It's the little details that make all the difference", he says. As in the image above, he photographs his creations with associated objects "to tell a story related to the book."

RIGHT "Book of Magic". This movie prop grimoire was inspired by the film *Willow*, a 1988 fantasy adventure directed by Ron Howard. The hero of the story is a farmer and student magician named Willow Ufgood, played by Warwick Davis. He gets embroiled in an epic struggle with the evil sorceress Queen Bavmorda. Good triumphs over evil, and at the end of the adventure Willow is gifted an ancient and powerful book of magic from one of his accomplices, the sorceress Fin Raziel. In historic terms, the angel Raziel, reputed for his knowledge of magic, was purportedly the author of a medieval Kabbalistic grimoire called the *Sefer Raziel HaMalakh*, which circulated from the thirteenth century.

ABOVE A representation of the *Necronomicon* portrayed in the *Evil Dead* trilogy of films (released between 1981 and 1992), directed by Sam Raimi. In the first film, five American college students stumble across the *Necronomicon*, described as bound in human flesh and "inked in human blood," in a creepy cabin deep in the woods. Alongside the book, they also find a tape recording of some of its incantations spoken by a deceased professor. When they play the tape, it unleashes a terrifying demon. The original book design was devised by special effects artist Tom Sullivan, and the demonic face on the cover has become iconic, inspiring many reproductions by fans.

RIGHT "Necronomicon (1)" by Matthew Corrigan, colored inks on paper. Over the last few decades a genre of "Lovecraftian art" has developed, with numerous artists creating either visual expressions from Lovecraft's words, or taking his ideas in novel artistic directions. The artwork can be "Lovecraftian" in feel and inspiration without explicitly depicting anything in Lovecraft's stories. An early example is the work of Swiss artist H. R. Giger (1940–2014) who called his 1977 book of esoteric fantasy art, *Necronomicon*. One of the illustrations in it, entitled, "Necronom IV," led to Giger being commissioned to design the monster in the film *Alien* (1979). As with the illustration here, the word "Necronomicon" becomes a Lovecraftian shorthand for his world and not just the title of Lovecraft's magic book.

LEFT AND ABOVE *Left.* "Necronomicon (2)" by Matthew Corrigan, colored inks on paper. *Above.* Image from a *Necronomicon* featured in the Lovecraftian horror film *Castle Freak* (2020). Both of these images are inspired by Lovecraft's god-monster Cthulhu, which he described as a huge malevolent beast with an octopus-like head, rubbery body, and a mass of feelers on its face. Cthulhu is imprisoned in an underwater city deep in the South Pacific and is both feared and worshipped by humans. Both illustrations also frame the monster with pseudoscripts and occult symbols, similar to the depictions of demons and their associated conjurations found on incantation bowls and in grimoires.

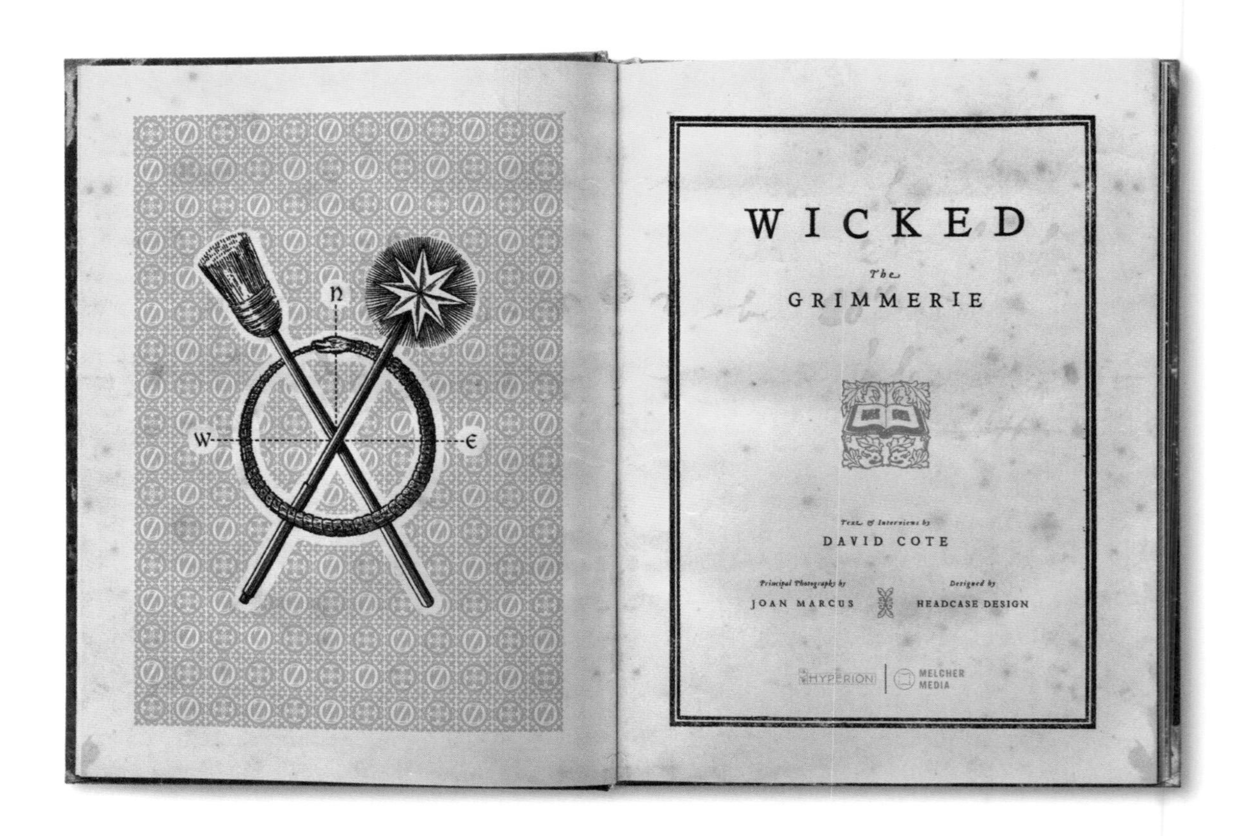

ABOVE Title page from *Wicked: The Grimmerie, a Behind-the-Scenes Look at the Hit Broadway Musical* by David Cote, Philadelphia (2005). This was a keepsake book for fans of the hit Broadway musical *Wicked*, which opened in 2003. The musical was inspired by Gregory Maguire's novel *Wicked: The Life and Times of the Wicked Witch of the West* (1995), which, in turn, was based on L. Frank Baum's novel *The Wonderful Wizard of Oz* (1900). The story concerns the friendship between Elphaba (the Wicked Witch of the West) and Glinda (the Good Witch). Elphaba is given a book of magic called *The Grimmerie*, which plays a significant role in subsequent *Wicked* novels written by Maguire. The designers of *Wicked: The Grimmerie* above, explained that they wanted it to look like an ancient magic book saying, "we juxtaposed fanciful Ozian illustrations with occult design elements, and painstakingly aged each page of the book."

RIGHT *June* was a British comic for girls that ran in various iterations from 1961–1974. One of its regular strips told the adventures of schoolgirl Vanessa from Venus, "the most amazing friend an Earth Girl ever had," who possesses "strange magical powers." As well as tapping into the 1960s popular interest in UFOs, the character also represents the emerging figure of the young, attractive female witch or magician that became popular at this time through the international success of the American television series *Bewitched* (1964–1972).

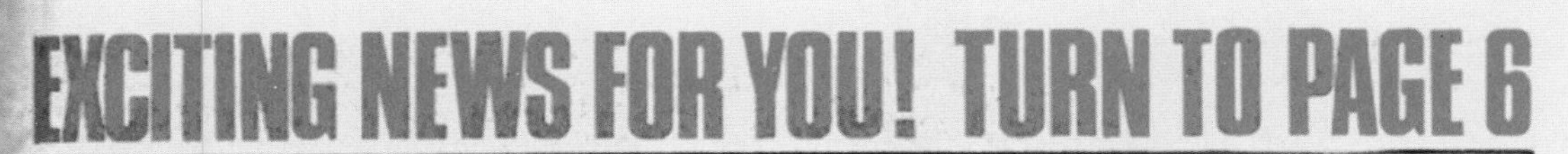

EXCITING NEWS FOR YOU! TURN TO PAGE 6
JUNE and SCHOOL FRIEND
13th JANUARY, 1968
EVERY TUESDAY
SEVENPENCE
VANESSA FROM VENUS
will put a magic spell on you to follow her adventures every week!
MAGIC SPELLS OF VENUS
MAGIC SPELLS OF VENUS
See the bright picture-story on page 24
Australia 10 cents ; New Zealand 10 cents ; South Africa 10 cents ; Rhodesia 1/- ; East Africa 1.00 cents ; West Africa 1/- ; Malaysia 40 cents ; Malta 9d. ;
Danmark Kr. 1.50 ; Deutschland Dm. .55 ; Espana Pes. 10.0 ; Italy Lire 120 ; Nederland Fl. .65 ; Norge Kr. 1.25 ; Portugal Esc. 5.0 ; Suomi Fm. .60 ; Sverige Kr. .75.

ABOVE AND RIGHT Images from Alexander D'Agostino, *The Fairy King's Grimoire*, printed on paper and leaves using solar reactive processes (2022). D'Agostino created this artwork during a Fellowship at the Folger Library, and was inspired by the library's late 16th-century grimoire with instructions for invoking spirits (see pages 142–143). *The Fairy King's Grimoire* also draws upon other magical material in the Folger collection, as well as a vintage gay porn collection found in a Baltimore antique shop. Both images depict Oberion/Oberon, the King of the Fairies, from medieval literature.

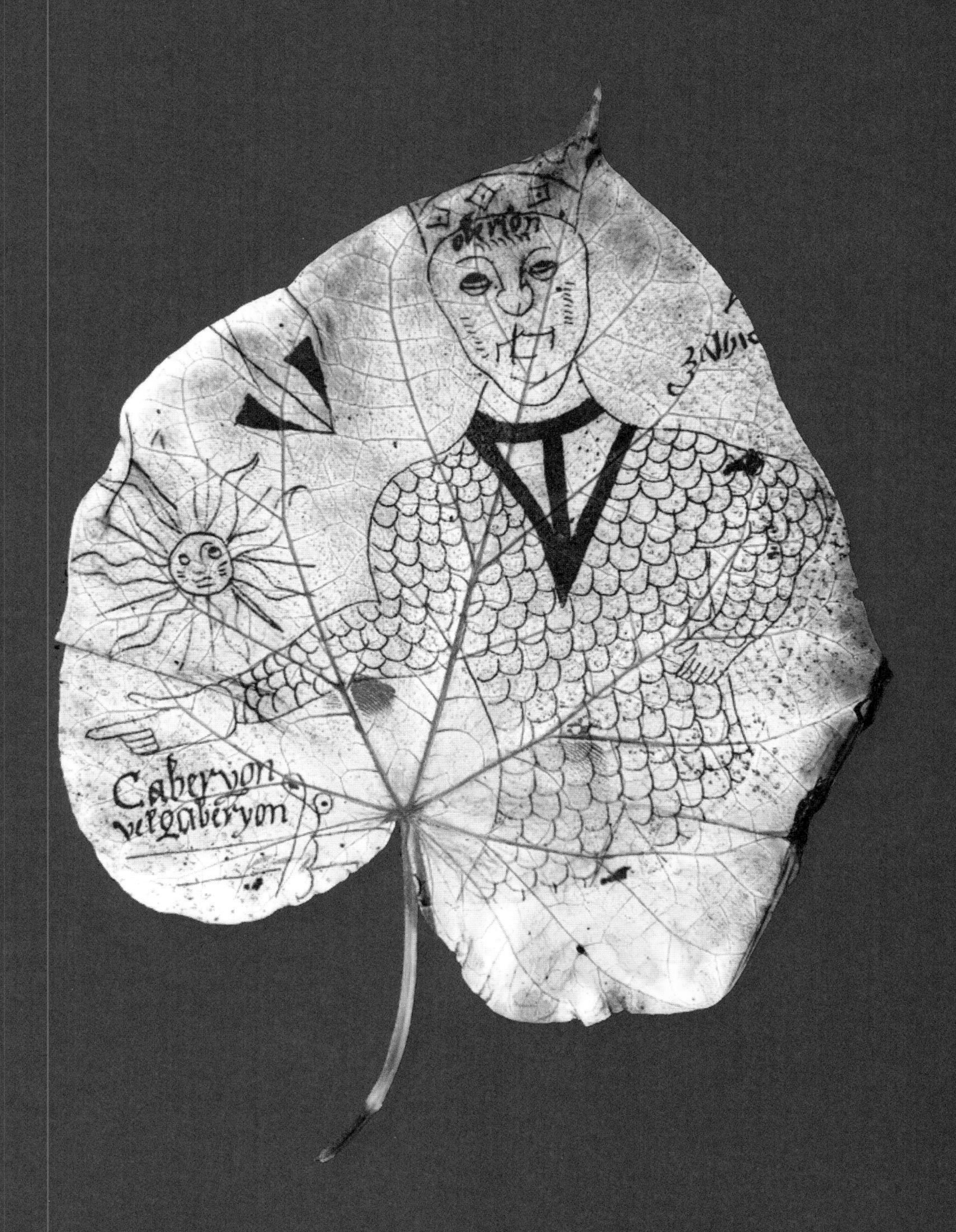
Caberyon
vel Gaberyon

Magic in Manga and Anime

The rich fantasy genre in Japanese manga comics and anime series demonstrates a remarkable fusion of western and eastern magical themes. Manga authors have long drawn upon the worlds of ancient Mediterranean history and European myth in time-travel stories. *Red River* (1995–2002), for instance, concerns a Japanese school student, who is magicked back to the ancient world by the curse of a Hittite empress. Many of the publications are largely unknown outside of Japan, and yet they contain numerous references to familiar characters and motifs from the history of magic. This melting pot tradition in manga owes much to the pioneering manga artist and historian Shigeru Mizuki (1922–2015), who cast his creative net wide and drew upon folklore, art, and magic traditions from beyond Japan. Judging from magical references in his work, he seems to have been familiar with Arthur Edward Waite's *Book of Black Magic and of Pacts* (1898). In a series of articles for a manga magazine in the 1960s, he plotted *yōkai* from around the world on a large map and included European variants such as Dracula, werewolves, wizards, and Medusa.

One of the most creative examples of this syncretic approach is *Rental Magica*, written by Makoto Sanda, which has appeared in manga and animation formats, and concerns the adventures of two rival magician societies, Astral and Goetia. As well as having Buddhist and Shinto masters in its team, one of Astral's members is an expert on witchcraft and druidic "Celtic magic," while another is a Germanic rune expert. The hereditary leader of Goetia, Adelicia Lenn Mathers (you get the name reference there), is a descendant of King Solomon and therefore can summon demons. Another character in the series is an English gentleman named MacGregor (Samuel Liddell Mathers added MacGregor to his name at one point), who teaches at a school for magicians and is an expert in Egyptian magic.

The possession of magic books is a strong theme in manga and anime, with one series even

LEFT The Grimoire Acceptance Ceremony in the Japanese manga *Black Clover* (see right). At the age of 15, trainee wizards are given their first grimoires in the great library of the Grimoire Tower. The allocated magic books begin to glow and float down from the shelves to land in the hands of their new masters.

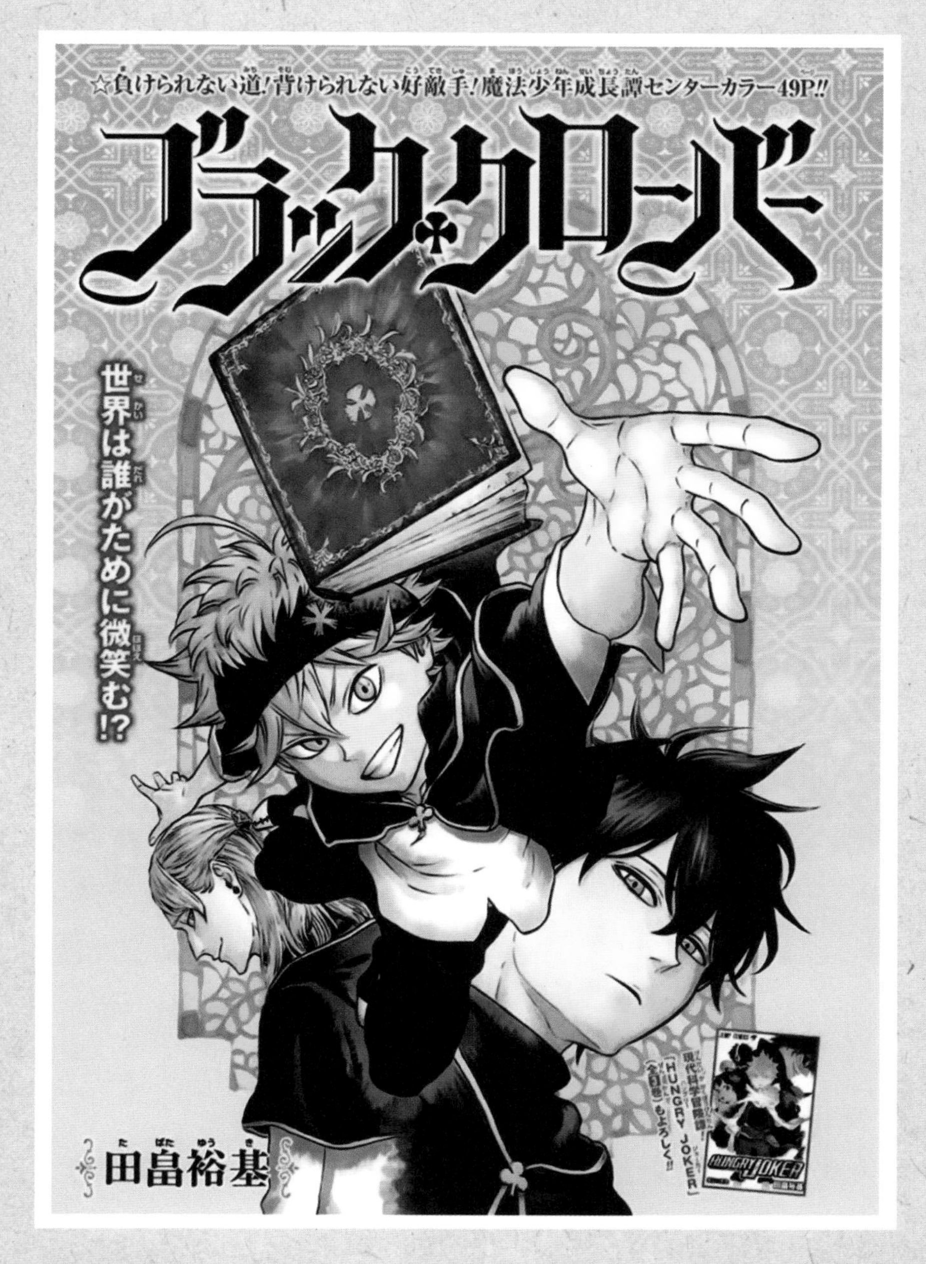

RIGHT Image from the Japanese manga *Black Clover* (2014), written and illustrated by Yūki Tabata. The story concerns two young orphans in the Clover Kingdom, named Asta and Yuno, whose quest is to work their way up the Order of the Magic Knights. Yuno is given a legendary four-leaf grimoire, once belonging to the kingdom's first Wizard King.

called *Grimoire of Zero* (2014–2017), in which the hero, Zero, is a child witch who teams up with a half-man, half-beast on a quest to find a fabled book of magic that could otherwise fall into evil hands. In other series, the esoteric library plays an important role in the aesthetics, as well as the plot. In the very successful *Cardcaptor Sakura* (1998), which developed a following beyond Japan thanks to an English dubbed version, ten-year-old Sakura finds the ancient "Book of Clow" in her dad's library, which contains a set of magical cards. She accidently sets the cards loose when the seal on the book is broken, and she then has a series of adventures to retrieve them. Clow Cards have generated their own fan-fiction designs, as well as commercial offerings.

SPElls
What does an artist
witch use in her Spells?
Water
fire
Air
Earth
ARt
mAgick

LEFT AND ABOVE Pages from *Artist Witch Spell Book* (2019) by Sarah Bellisario, pen and inks, and letter transfers. The artwork was created for "The Sketchbook Project," and was housed at the Brooklyn Art Library in New York. Working with the theme of "Legacy," and exploring the twin notions of both "Artist" and "Witchcraft," Bellisario's aim as a practicing witch was "to create an informative but tongue-in-cheek collection of work about what it means to be an 'Art Witch'." The collected work explores themes including the ephemera of witchcraft, ritual and spell work, and the artist herself as a magical creator.

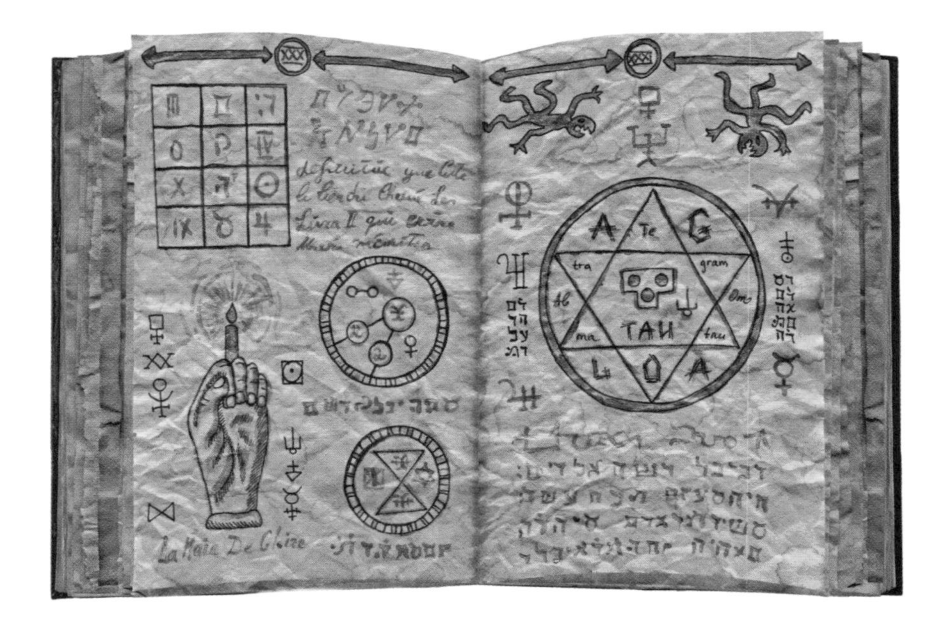

ABOVE Images from an "Old magic book with magic mystic symbols and drawings" by Vera Petruk, watercolor and collage. Petruk's work draws inspiration from eastern and western esoteric and occult themes. These illustrations show several pages that Petruk has created for an imagined European grimoire. They play with Hebraic-looking script, the characters or signs of spirits, magic squares, demon images, and talismans. She has also included an example of the Hand of Glory, which is an iconic image originally from the *Petit Albert* (see page 173), which was a very popular 18th- to 19th-century French book of spells, recipes, and household tips. The Hand was that of an executed criminal, and, when preserved, if a candle was placed in it and lit, it would stupefy the inhabitants and so allow burglars to go about their business.

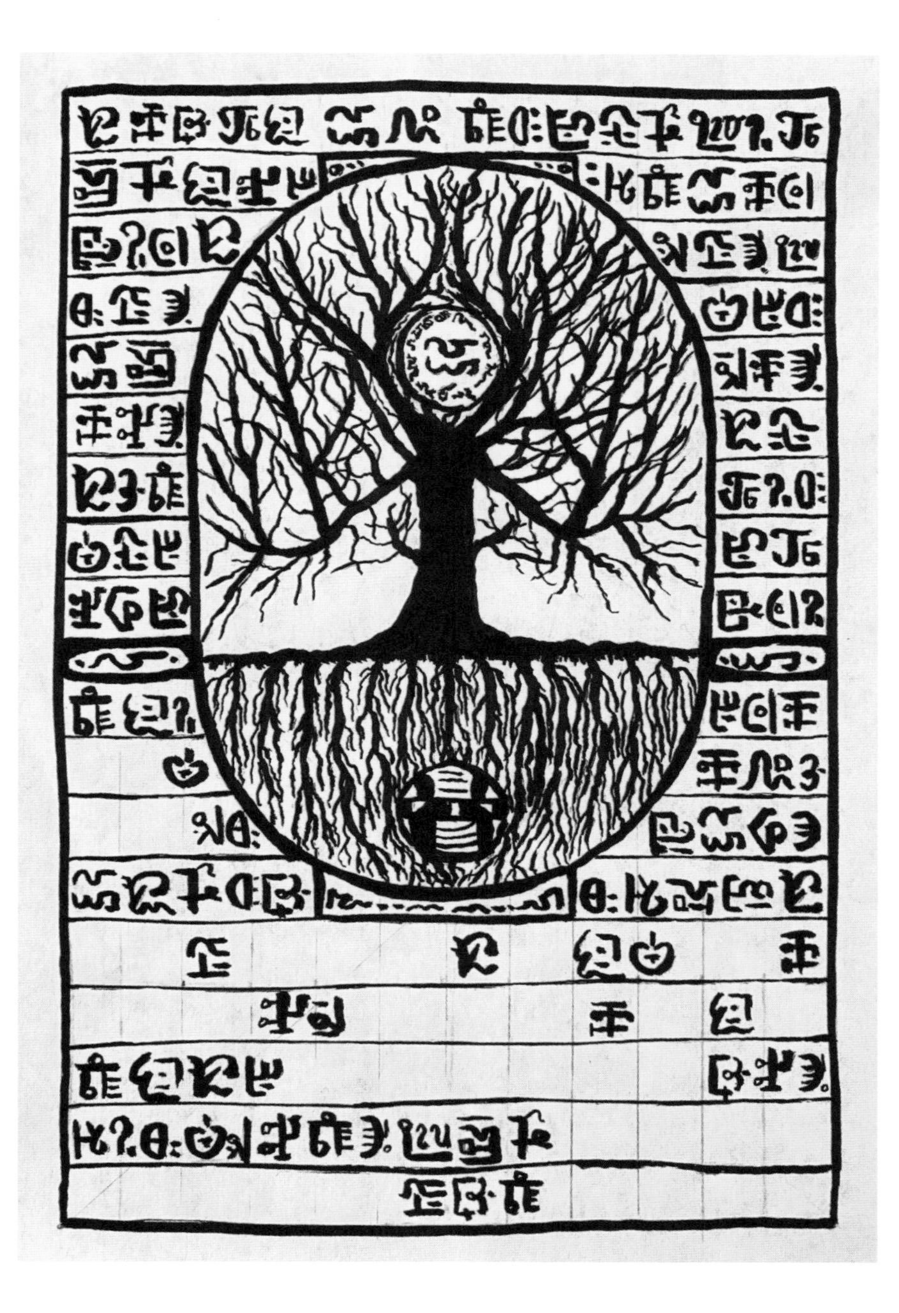

ABOVE "Grimoire #2," a drawing by American artist Jim Ingram, ink on silk. This plays on the imagery of pseudo-scripts combined with the tree as a symbol of life. With its simple use of blank ink, the drawing has the feel of the early woodblock print magic texts of medieval China and the Arabic *tarsh* talismans (see page 90).

ABOVE AND RIGHT Pages from *The Book of Shadows, the Lost Code of the Tarot* by Andrea Aste, in red and black inks on artificially aged paper (2015). Aste's artwork is part of a multimedia arts project based on the question, "Who invented the tarot?" The content and feel of her *Book of Shadows* drew inspiration from Renaissance esoteric works, including Arabic geometry and magic squares, and the oldest surviving tarot decks. She says, "I have wanted to echo a sense of magic and mystery typical of a time where ancient myth and new ideas were fused."

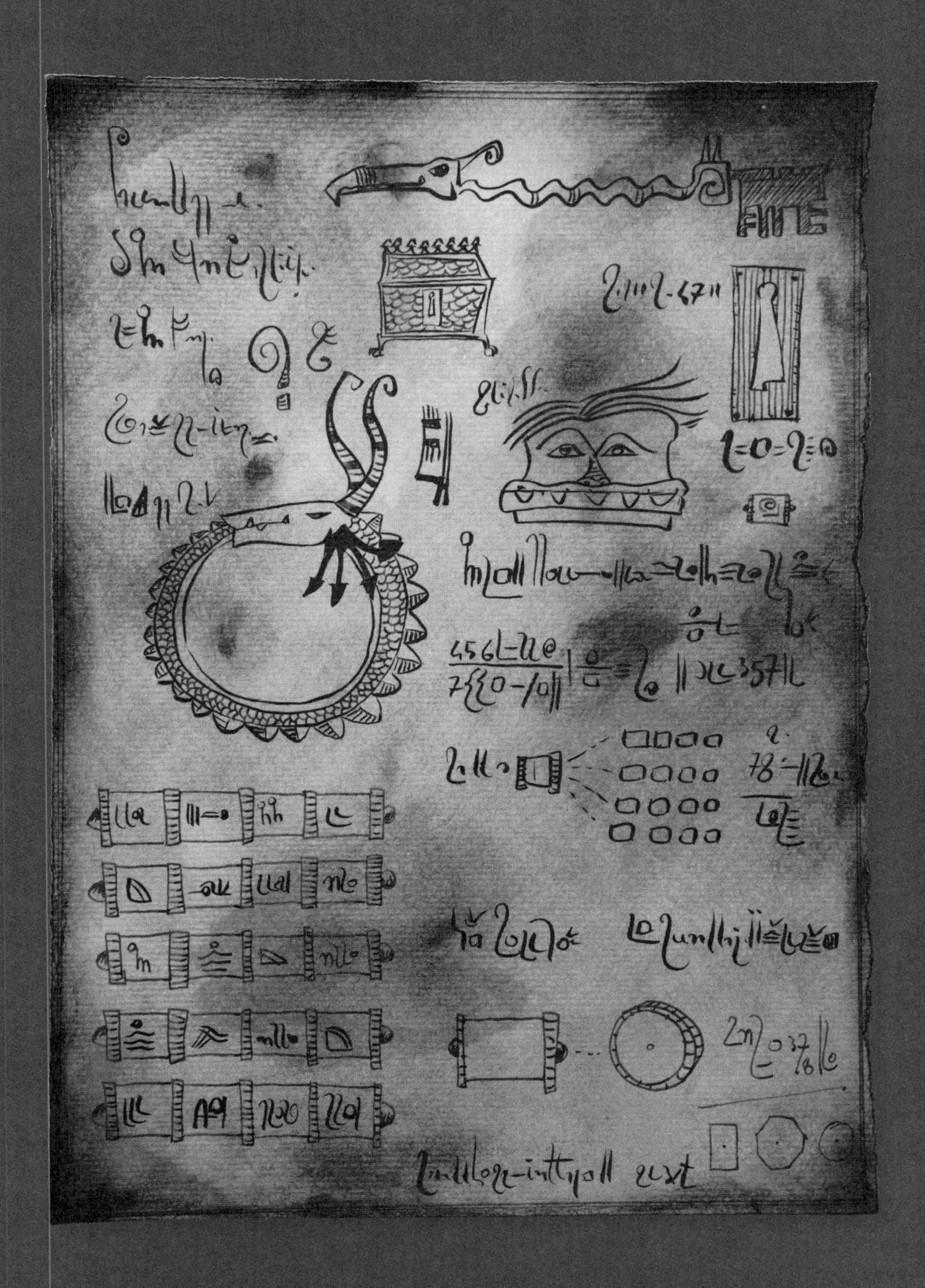

LEFT AND ABOVE Watercolor illustrations of jinn from an early 20th-century copy of the influential medieval book, *Kitāb-i'Ajā'ib-i makhlūqāt* (*Book of The Wonders of Creatures*), Isfahan, Iran (c. 1906–1921). Originally compiled by the Iranian scholar Zakariyya al-Qazwini (d. 1283), this work of cosmography, astrology, and magic has a rich history of illustrated manuscript copying. That such sumptuously painted examples were still being produced over 600 years later demonstrates the continued vibrancy of artistic manuscript magic in non-Western occult traditions.

Bibliography

Broad Overviews

Bailey, Michael D., *Magic and Superstition in Europe: A Concise History from Antiquity to the Present* (Lanham, 2007).

Boschung, Dietrich, and Jan N. Bremmer (eds), *The Materiality of Magic* (Paderborn, 2015).

Collins, David J., *Magic and Witchcraft in the West: From Antiquity to the Present* (Cambridge, 2015).

Copenhaver, Brian, *The Book of Magic from Antiquity to the Enlightenment* (London, 2015).

Davies, Owen (ed.), *The Oxford Illustrated History of Witchcraft and Magic* (Oxford, 2017).

Davies, Owen, *Magic: A Very Short Introduction* (Oxford, 2012).

Davies, Owen, *Grimoires: A History of Magic Books* (Oxford, 2009)

Gosden, Chris, *The History of Magic: From Alchemy to Witchcraft, from the Ice Age to the Present* (London, 2021).

Hutton, Ronald, *The Witch: A History of Fear, from Ancient Times to the Present* (London, 2017).

Otto, Bernd-Christian. 2016. "Historicising 'Western Learned Magic': Preliminary Remarks." *Aries* 16: 161–240

Chapter 1

Betz, Hans Dieter (ed.), *The Greek Magical Papyri in Translation: Including the Demotic Spells*, 2nd ed. (1992).

Bohak, Gideon. (2007) *Ancient Jewish Magic: A History* (Cambridge, 2007).

Bréard, Andrea, "Cracking bones and numbers: solving the enigma of numerical sequences on ancient Chinese artifacts," *Archive for History of Exact Sciences* 74 (2020), 313–43.

Dasen, Véronique, "Sexe et sexualité des pierres dans l'Antiquité gréco-romaine," in Véronique Dasen et Jean-Michel Spieser (eds), *Les savoirs magiques et leur transmission de l'Antiquité à la Renaissance* (Florence, 2014), 195–220.

Dieleman, Jacco, *Priests, Tongues, and Rites: The London-Leiden Magical Manuscripts and Translation in Egyptian Ritual (100-300 CE)* (Leiden 2005).

Faraone, Christopher A. and Dirk Obbink (eds), *Magika Hiera: Ancient Greek Magic & Religion* (Oxford, 1991).

Faraone, Christopher A., "A Greek Magical Gemstone from the Black Sea. Amulet or Miniature Handbook," *Kernos* 23 (2010), 91–114.

Frankfurter, David (ed.) *Guide to the Study of Ancient Magic* (Leiden, 2019).

Guozhang, Liu, *Introduction to the Tsinghua Bamboo-Strip Manuscripts*, trans Christopher Foster and William N. French (Brill: Leiden, 2015).

Harper, Donald and Marc Kalinowski (eds), *Books of Fate and Popular Culture in Early China: The Daybook Manuscripts of the Warring States, Qin, and Han* (Leiden, 2017).

Harper, Donald, "The Textual Form of Knowledge: Occult Miscellanies in Ancient and Medieval Chinese Manuscripts, Fourth Century B.C. to Tenth Century A.D.," in Florence Bretelle-Establet (ed.), *Looking at it from Asia: The Processes that Shaped the Sources of History of Science* (Dordrecht, 2010), pp. 37–80.

Johnston, Jay and Iain Gardner (eds), *Drawing Spirit: The Role of Images and Design in the Magical Practice of Late Antiquity* (Berlin, 2022).

Kitz, Anne Marie, *Cursed Are You!: The Phenomenology of Cursing in Cuneiform and Hebrew Texts* (Winona Lake, 2014).

Kiyanrad, Sarah, Christoffer Theis, and Laura Willer (eds.), *Bild und Schrift auf "magischen" Artefakten* (Berlin 2018).

McKie, Stuart, *Living and Cursing in the Roman West: Curse Tablets and Society* (London, 2022).

Meyer, Marvin and Richard Smith (eds), *Ancient Christian magic: Coptic Texts of Ritual Power* (San Francisco, 1994).

Müller-Kessler, C., "Interrelations between Mandaic lead rolls and incantation bowls," in Tzvi Abusch and Karel van der Toorn (eds), *Mesopotamian Magic: Textual, Historical and Interpretive Perspectives* (Leiden, 2000), pp. 197–209.

Rebiger, Bill, "'Write on Three Ribs of a Sheep': Writing Materials in Ancient and Mediaeval Jewish Magic," in Irina Wandrey (ed.), *Jewish Manuscript Cultures: New Perspectives* (Berlin, 2017), pp. 339–59.

Sanzo, Joseph E., "At the Crossroads of Ritual Practice and Anti-Magical Discourse in Late Antiquity: Taxonomies of Licit and Illicit Rituals in Leiden, Ms. AMS 9 and Related Sources," *Magic, Ritual, and Witchcraft* 14 (2019), 230–54.

Chapter 2

Boudet, Jean-Patrice, "Les who's who démonologiques de la Renaissance et leurs ancêtres médiévaux." *Médiévales* 44 (2003), 117–140.

Stefano Carboni, "The 'Book of Surprises' (*Kitab al-bulhan*) of the Bodleian Library," *The La Trobe Journal* 91 (2013), 22–34.

Harari, Yuval, *Jewish Magic before the Rise of Kabbalah* (Detroit, 2017).

Holsinger, Bruce, *On Parchment: Animals, Archives, and the Making of Culture from Herodotus to the Digital Age* (New Haven, 2022).

Kieckhefer, Richard, *Magic in the Middle Ages* (Cambridge, 1989).

Kim, Sujung, "A Star God Is Born: *Chintaku Reifujin* Talismans in Japanese Religions," *DePauw University Religious Studies Faculty publications* 13 (2022), 1–17.

Klaassen, Frank, *The Transformations of Magic: Illicit Learned Magic in the Later Middle Ages* (University Parks, 2013).

Mollier, Christine, "Astrological Talismans and Paper Amulets from Dunhuang: Typology and Function," *Dunhuang Turfan Studies* 15 (2015), 505–19.

Mollier, Christine, "La Méthode de l'Empereur du Nord du mont Fengdu: Une tradition exorciste du taoïsme médiéval," *T'oung Pao*/通報 83 (1997), 329–85.

Page, Sophie and Catherine Rider (eds), *The Routledge History of Medieval Magic* (London, 2021).

Page, Sophie, *Magic in Medieval Manuscripts* (London, 2004).

Page, Sophie, *Astrology in Medieval Manuscripts* (London, 2002).

Porter, Venetia, Liana Saif, and Emilie Savage-Smith, "Medieval Islamic Amulets, Talismans, and Magic," in Finbarr Barry Flood and Gülru Necipoğlu (eds), *A Companion to Islamic Art and Architecture* (Hoboken 2017), pp. 521–57.

Robson, James, "Signs of Power: Talismanic Writing in Chinese Buddhism," *History of Religions* 48 (2008), 130–69.

Ryder, Catherine, *Magic and Religion in Medieval England* (London, 2012).

Saif, Liana, *The Arabic Influences on Early Modern Occult Philosophy* (Basingstoke, 2015).

Savage-Smith, Emilie (ed.), *Magic and Divination in Early Islam* (Aldershot, 2014).

Schaik, Sam Van, *Buddhist Magic: Divination, Healing, and Enchantment Through the Ages* (Boulder, 2020).

Segol, Marla, *Word and Image in Medieval Kabbalah: The Texts, Commentaries, and Diagrams of the Sefer Yetsirah* (New York, 2012).

Skemer, Don C., *Binding Words: Textual Amulets in the Middle Ages* (University Park, 2006).

Zadeh, Travis, "Commanding Demons and Jinn: The Sorcerer in Early Islamic Thought," in Alireza Korangy and Dan Sheffield (eds), *No Tapping around Philology: A Festschrift in Honor of Wheeler McIntosh Thackston Jr.'s 70th Birthday* (Wiesbaden 2014), 131–60.

Chapter 3

Adams, Alison and Stanton J. Linden (eds), *Emblems and Alchemy* (Glasgow, 1998).

Almond, Philip C., *England's First Demonologist: Reginald Scot and "the Discoverie of Witchcraft"* (London 2014).

Balbale, Abigail Krasner, "Magical words: Arabic amulets in Christian Spain," in Suzanna Ivanič, Mary Laven, and Andrew Morrall (eds), *Religious Materiality in the Early Modern World* (Amsterdam, 2020), pp. 211–28.

Barbierato, Federico, *The Inquisitor in the Hat Shop: Inquisition, Forbidden Books and Unbelief in Early Modern Venice* (Farnham, 2012).

Barbierato, Federico, "Ripped Shoes and Books of Magic: Practice and Limits of Inquisitorial Control on the Circulation of Forbidden Books in Venice between the Sixteenth and Seventeenth Centuries," in Katherine Aron-Beller and Christopher Black (eds), *The Roman Inquisition: Centre versus Peripheries* (Leiden, 2018), pp. 207–233.

Brock, Michelle D., Richard Raiswell, and David R. Winter (eds), *Knowing Demons, Knowing Spirits in the Early Modern Period* (Cham, 2018).

Buringh, Eltjo and Jan Luiten Van Zanden, "Charting the 'Rise of the West': Manuscripts and Printed Books in Europe: A Long-Term Perspective from the Sixth through Eighteenth Centuries," *The Journal of Economic History* 69 (2009), 409–45.

Davies, Owen, *Cunning-Folk: Popular Magic in English History* (London 2003).

Davies, S.F., "The Reception of Reginald Scot's Discovery of Witchcraft: Witchcraft, Magic, and Radical Religion," *Journal of the History of Ideas* 74 (2013), 381–401.

Eamon, William, *Science and the Secrets of Nature: Books of Secrets in Medieval and Early Modern Culture* (Princeton, 1994).

Forshaw, Peter J., "Alchemical Images," in D. Jalobeanu and C. T. Wolfe (eds), *Encyclopedia of Early Modern Philosophy and the Sciences* (Cham 2020), pp. 1–10.

Gordon, Stephen. "Necromancy for the Masses? A Printed Version of the *Compendium magiae innaturalis nigrae*," *Magic, Ritual, and Witchcraft* 13 (2018), 340–80.

Kassell, Lauren, "Secrets Revealed: Alchemical Books in Early-Modern England," *History of Science* 49 (2011), 61–87.

Schaefer, Karl R., *Enigmatic Charms: Medieval Arabic Block Printed Amulets in American and European Libraries and Museums* (Leiden, 2006).

Chapter 4

Bellingradt, Daniel and Bernd-Christian Otto, *Magical Manuscripts in Early Modern Europe: The Clandestine Trade in Illegal Book Collections* (Cham, 2017).

Burtea, Bogdan, "Traditional Medicine and Magic According to Some Ethiopian Manuscripts from European Collections," *Aethiopica* 18 (2015), 87–100.

Bühnemann, Gudrun, *Mandalas and Yantras in the Hindu Traditions* (Leiden, 2003).

Chernetsov, Sevir B., "Ethiopian Magic Literature," *Scrinium* (2006), 92–113.

Chirapravati, Pattaratorn (ed.), *Divination au royaume de Siam: Le corps, la guerre, le destin* (Paris 2011).

Davies, Owen, "Narratives of the Witch, the Magician, and the Devil in Early Modern Grimoires," in Bernd-Christian Otto and Dirk Johannsen (eds), *Fictional Practice: Magic, Narration, and the Power of Imagination* (Leiden 2021), pp. 91–109.

Donmoyer, Patrick J., *Powwowing in Pennsylvania: Braucherei and the Ritual of Everyday Life* (Pennsburg, 2018).

Engmann, Rachel Ama Asaa, "Mysticism, Enchantment and Charisma Reincarnated: Nineteenth-Century Islamic Talismans, Vernacularization and Heritage Values in Contemporary Asante," *Material Religion* 15 (2019), 221–39.

Fani, Sara, "Magic, traditional medicine and theurgy in Arabo-Islamic manuscripts of the Horn of Africa: a brief note on their description," in Alessandro Bausi et al (ed.), *Essays in Ethiopian Manuscript Studies: Proceedings of the International Conference Manuscripts and Texts, Languages and Contexts: The Transmission of Knowledge in the Horn of Africa* (Wiesbaden 2015), pp. 273–80.

Foster, Michael Dylan, *Pandemonium and Parade: Japanese Monsters and the Culture of Yōkai* (Berkeley 2009).

Graham, Lloyd D., "The Magic Symbol Repertoire of Talismanic Rings from East and West Africa," *Humanities Commons* (2020).

Klaassen, Frank, and Sharon Hubbs Wright, *The Magic of Rogues: Necromancers in Early Tudor England* (University Park 2021).

Klaassen, Frank, *Making Magic in Elizabethan England: Two Early Modern Vernacular Books of Magic* (University Park, 2019).

Lillehoj, Elizabeth, "Transfiguration: Man-Made Objects as Demons in Japanese Scrolls," *Asian Folklore Studies* 54 (1995), 7–34.

Moriggi, Marco and Siam Bhayro (eds), *Studies in the Syriac Magical Traditions* (Leiden, 2022).

Ohrvik, Ane, *Medicine, Magic and Art in Early Modern Norway: Conceptualizing Knowledge* (London, 2018).

Papp, Zília, *Anime and Its Roots in Early Japanese Monster Art* (Folkestone 2010).

Peterson, Joseph H., James R. Clark, and Dan Harms (eds), *The Book of Oberon: A Sourcebook of Elizabethan Magic* (Woodbury, 2015).

Pontzen, Benedikt, "What's (not) in a leather pouch? Tracing Islamic amulets in Asante, Ghana," *Africa* 90 (2020), 870–89.

Probert, Marcela A. Garcia and Petra M. Sijpesteijn (eds), *Amulets and Talismans of the Middle East and North Africa in Context* (Leiden, 2022).

Owusu-Ansah, David, *Islamic Talismanic Tradition in Nineteenth-Century Asante* (Lewiston, 1991).

Regourd, Anne, "A Twentieth-Century Manuscript of the *Kitāb al-Mandal al-sulaymānī* (ies Ar. 286, Addis Ababa, Ethiopia): Texts on Practices & Texts in Practices," in Marcela A. Garcia Probert and Petra M. Sijpesteijn (eds), *Amulets and Talismans of the Middle East and North Africa in Context* (Leiden, 2022), pp. 47–77.

Reider, Noriko T., *Japanese Demon Lore: Oni, from Ancient Times to the Present* (Logan, 2010).

Satterley, Renae, "Robert Ashley and the Authorship of Newberry MS 5017, The Book of Magical Charms," *Manuscript Studies* 6 (2022), 3–34.

Teygeler, René, "*Pustaha*: A Study into the Production Process of the Batak Book," *Manuscripts of Indonesia* 149 (1993), 593–611.

Westerkamp, Willem, "From *Singa* to *Naga Padoha*: The Making of a Magical Creature", *Indonesia and the Malay World* 37 (2009), 163–81.

Yahya, Farouk, *Magic and Divination in Malay Illustrated Manuscripts* (Leiden, 2015).

Chapter 5

Barreto, Inês Teixeira, "Da Mandinga à Macumba: a trajetória do Livro de São Cipriano no Brasil," PhD thesis, Pontifícia Universidade Católica de São Paulo, 2022.

Bohak, Gideon, "How Jewish Magic Survived the Disenchantment of the World," *Aries–Journal for the Study of Western Esotericism* 19 (2019), 7–37.

Castro, Félix Francisco, "El Libro de San Cipriano," *Hibris: Revista de bibliofilia* 27 (2005), 15–25.

Davies, Owen, *Grimoires: A History of Magic Books* (Oxford, 2009).

Gellman, Uriel, "Popular Religion and Modernity: Jewish Magic Books in Eastern Europe in the Nineteenth Century," *Polin Studies in Polish Jewry* 33 (2021), 185–202.

Junqueira, Luis Fernando Bernardi, "Revealing Secrets: Talismans, Healthcare and the Market of the Occult in Early Twentieth-century China," *Social History of Medicine* 34 (2021), 1068–1093.

Krögel, Alison, "Asustar para enseñar: el papel didáctico de los personajes malévolos en la narrativa oral quechua," *Bolivian Research Review/Revista de Estudios Bolivianos* 8 (2009), 2–26.

Leitão, J. V., "The Folk and Oral Roots of the Portuguese «Livro de São Cipriano»," *International Journal of Heritage and Sustainable Development* 4 (2015), 129-39.

Long, Carolyn Morrow, *Spiritual Merchants: Religion, Magic, and Commerce* (Knoxville, 2001).

Mukharji, Projit Bihari, "Magic of Business: Occult Forces in the Bazaar Economy," in Ajay Gandhi, Barbara Harriss-White, Douglas E. Haynes, and Sebastian Schwecke (eds), *Rethinking Markets in Modern India: Embedded Exchange and Contested Jurisdiction* (Cambridge, 2020), pp. 85-116. My thanks to Projit for also sending me images of his collection of South Asian magic books.

Page, Hugh R. Jr., "Post-Imperial Appropriation of Text, Tradition, and Ritual in the Pseudonymous Writings of Henri Gamache," in Stephen Finley, Margarita Guillory, and Hugh Page Jr (eds), *Esotericism in African American Religious Experience* (Leiden, 2015), 152–61.

Rockwell, Elsie, "Relaciones con la cultura escrita en una comunidad nahua a principios del siglo XX: temas recurrentes en los relatos orales," in Antonio Castillo Gómez and Verónica Sierra Blas (eds), *Senderos de ilusión. Lecturas populares en Europa y América latina (Del siglo XVI a nuestros días)* (Gijón 2007), pp. 259–79.

Schindler, Helmut and Franz Xaver Faust, "Relaciones interétnicas de los curanderos en el suroccidente colombiano," *Anthropologica del Departamento de Ciencias Sociales* 18 (2000), 281–94.

Selove, Emile, "Magic as Poetry, Poetry as Magic: A Fragment of Arabic Spells," *Magic, Ritual, and Witchcraft* 15 (2020), 33–57.

Chapter 6

Brick, Emily (2018) "Spellbound: The Significance of Spellbooks in the depiction of Witchcraft on Screen," in *Terrifying Texts: Essays on Books of Goodand Evil in Horror Cinema*.

Grill, Genese, "Almandal Grimoire: The Book as Magical Object," *The Georgia Review* 69 (2015), 514–41.

Hutton, McWilliams, Stuart, "Aleister Crowley's Graphomania and the Transformations of Magical Inscriptivity," *Preternature* 5 (2016) 59–85.

Shamoon, Deborah, "The *Yōkai* in the Database: Supernatural Creatures and Folklore in Manga and Anime," *Marvels & Tales* 27 (2013), 120–33.

M. E. Warlick, M.E., "Leonora Carrington's Esoteric Symbols and their Sources", *Studia Hermetica Journal* 1 (2017), 56–83.

White, Ethan Doyle, *Wicca: History, Belief and Community in Modern Pagan Witchcraft* (Brighton, 2016).

Index

Page numbers in *italic* type refer to illustrations or their captions.

N

O

P

Q

R

S

T

Image credits

Images on the pages listed below are reproduced by kind permission of the owners. The publishers have made every effort to contact the owners and apologise for any unwitting infringement. Specific acknowledgements are as follows:

Alexander D'Agostino. Images from Folger Shakespeare Library Fellowship Project: 'The Fairy King's Grimoire': **236**, **237**. © Alamy: **76** (both) (Bodleian Libraries, University of Oxford/The Picture Art Collection); **101** (Heritage Image Partnership Ltd); **128** (© Fine Art Images/ Heritage Images); **216** (Photo 12); **217** (Collection Christophel); **222** (Charles Walker Collection); **229** (Chris Howes/Wild Places Photography); **231**, **232** (Matthew Corrigan); **233** (© RLJE Films /Courtesy Everett Collection); **242** (bottom) (Vera Petruk). © Amgueddfa Cymru—National Museum Wales: **37**. Andrea Aste, Courtesy of, www.andreaaste.co.uk: **244**, **245**. Andy Polaine, Courtesy of, (BY-SA 4.0 CC Licence): **29**. Apostolic Library, Vatican City © Vatican Library: **60**. Archaeological Museum of Pella, by permission of the Ephorate of Antiquities of Pella, Ministry of Culture and Sports /Archaeological Resources Fund: **36**. Art Institute of Chicago, Lucy Maud Buckingham Collection (CC0): **58**. Author's collection: **93**, **112** (both), **113**, **172**, **176**, **183**, **184** (both), **185**, **188**, **187**, **192**, **194–199**, **200–207**. Bayerische Staatsbibliothek München, Res/4 Crim. 15a, title page: **92**. Beinecke Rare Book and Manuscript Library, General Collection, Yale University: **69** (both), **83** (both), **168–169**. Biblioteca Pública Piloto de Medellín: **182**. © Bibliothèque nationale de France, Paris: **67** (bottom), **72** (left), **75**, **86–87**, **119**, **166**, **220**, **221**. BLACK CLOVER © 2016 by Yuki Tabata/SHUEISHA Inc: **238**, **239**. Bodleian Library, University of Oxford: **74** (MS. Bodl. Or.133, fo.27b); **84** (MS. Rawl. D.252, 28v, 29r); **85** (MS. Rawl. D.252, fo.104v); **144** (MS. Add. A.287, fo.137v); **145** (MS. Ashmole 972). Bologna, Museo Civico Archeologico; Photo: Celia Sánchez Natalías: **17**. bpk / Vorderasiatisches Museum, SMB / Olaf M.Teßmer: **33**. Bridgeman Images, London: **19** (© Brooklyn Museum of Art); **20**, **21**, **46**, **54**, **55**, **56**, **59**, **71**, **72** (right), **73**, **78**, **80**, **81**, **82** (both, and cover, background), **86** (left), **118**, **132**, **133**, **139**, **160**, **164**, **165** (© British Library Board. All Rights Reserved); **70** (Bibliothèque nationale de France, Paris); **212** (Private collection); **235** (© The Advertising Archives). The British Museum, © Trustees of: **16**, **22**, **43** (both), **44–45**, **52–53**, **62**, **63**. Buratti Art Group, Image courtesy: **213**. Cambridge University Library, Reproduced by kind permission of the Syndics of: **32**, **64**, **65**. Daniel W. VanArsdale collection: **189** (top). © The David Collection, Copenhagen, 85–2003; Photo Pernille Klemp (museum scan 2016): **99**. © The Doreen Valiente Foundation. All rights reserved: **214**. Dorotheum Vienna, auction catalogue 28.05.2016: **166**, **167**. Durga Pustak Bhandar: **179**. Editorial Caymmi, Bueonos Aires: **177**, **178**. Folger Shakespeare Library (CC BY-SA 4.0): **91**, **142**, **143**. Fondation Martin Bodmer, Geneva: **154**, **155**. Franklin and Marshall College, Lancaster, PA, Courtesy of Archives and Special Collections: **170–171**. Gallery Wendi Norris, San Francisco, Courtesy of; © Estate of Leonora Carrington / ARS, NY and DACS, London 2023. German Book and Type Museum of the German National Library Leipzig, Cultural History Collection, Signature: KHS EO 1960/002: **34**. Getty Research Institute, Los Angeles (PDM): **122**, **123**. Gonnelli Auction House, Florence, Courtesy of: **129**. Heilman Collection of Patrick J. Donmoyer: **162**, **163**, **190–191**. Honolulu Museum of Art. Collections of the, Gift of Mr. Edwin Binney, III, 1972 (4104.1): **79**. J. Paul Getty Museum, The, Villa Collection, Malibu; Digital image courtesy of Getty's Open Content Program: **18**. Jagiellonian Library, Jagiellonian University, Krakow: **77**. Jim Ingram, frontis: **210–211**, **243**. Kelsey Museum of Archaeology, University of Michigan: KM 26054 (both sides) Amulet with Ouroboros and cock-headed anguipede, from Egypt, AD 100–500. Green jasper, 37 x 27 x 4 mm): **39**. © Klassik Stiftung Weimar,

Herzogin Anna Amalia Bibliothek: **151** (both). Leipzig University Library: **134, 152, 153**. Library of Congress, Washington D.C. (PDM): **148, 149** (both; World Digital Library); **193**. Los Angeles County Museum of Art (PDM): **90**. Maia Yasmin Gala Benedini/Designer: **215**. Mark Nattkemper, courtesy of: **230**. Metropolitan Museum of Art, New York (OA), The: **14** (86.11.64; Purchase, 1886); **24** (86.11.130; Purchase, 1886); **25** (44.4.54; Rogers Fund, 1944); **26** (50.85; Fletcher Fund, 1950); **98** (1978.546.32; Gift of Nelly, Violet and Elie Abemayor, in memory of Michel Abemayor, 1978); **135** (95.66; Museum Purchase, transferred from the Library); **138** (1998.199; Purchase, Friends of Islamic Art Gifts, 1998.); **167** (both; 2012.5; Purchase, Marie Sussek Gift, 2012). MilleCuirs – Etienne Millete www.etsy.com/millecuirs, courtesy of: **228**. Museum of Cultures of the World, Institut de Cultura de Barcelona: **161**. © The Museum of Witchcraft and Magic, Boscastle: **223**. National and University Library of Iceland, The: **136, 146, 147**. National Archives of Japan, Tokyo (CC-BY 4.0): **149** (bottom). National Central Library of Rome (PDM): **94, 110, 111**. National Library of Medicine, Bethesda, Maryland (PDM): **186**. National Library of Israel, courtesy of Ktiv Project: **15**. National Library of Malaysia, Courtesy of the: **11**. National Library of Norway, Oslo: **156, 157**. National Library of Portugal (PDM): **173**. National Museum of Antiquities, Leiden (CC0), AMS 9: **40, 51**. National Museum of Asian Art, Smithsonian Institution, Arthur M. Sackler Collection, The Dr. Paul Singer Collection of Chinese Art of the Arthur M. Sackler Gallery, Smithsonian Institution; a joint gift of the Arthur M. Sackler Foundation, Paul Singer, the AMS Foundation for the Arts, Sciences, and Humanities, and the Children of Arthur M. Sackler, S2012.9.461: **31**. National Museum of Asian Art, Smithsonian Institution, Freer Collection, Gift of Charles Lang Freer, F1908.45.12 and F1908.45.13 (detail): **41, 68**. Newberry Digital Collections (Newberry Library), Chicago: **140** (both). Norwegian Folklore Archives, University of Oslo; NFS Joh. Olsen: **158, 159**. Open source image: **175, 180, 181, 208, 209**. Penn Museum, Philadelphia, Courtesy of the (object # 29-20-1; B2945; E2775E): **12–13, 27, 30, 42**. Photo: F. Bucher. (CC-BY-SA-3.0): **66**. Photo: Karen Roe, 2012 (CC BY 2.0): **218**. Photo by Bruce and Kenneth Zuckerman, West Semitic Research, Courtesy Private Collection: **38**. Photo: Zh, via Wikipedia (CC BY-SA 3.0): **35**. Princeton University Library (PDM): **246, 247**. Princeton University Library, Cotsen Memorial Library (Cruik 18-.3): **103**. Private collection (PDM): **114, 115** (both), **225**. Rijksmuseum, Amsterdam (PDM): **104–105**. Royal Library, National Library of the Netherlands, Leiden: **93**. Royal Zoological Society Natura Artis Magistra/ National Museum of World Cultures. Coll.no. TM-A-1389: **9, 10**. ©sarahbellisario: **240, 241**. Saint John's University and the Hill Museum & Manuscript Library in Collegeville, Minnesota, Image courtesy of. Published with permission of the owners. All rights reserved: **7**. Sigil Engine, Courtesy of; www.sigilengine.com: **226, 227** (both). Shutterstock/Vera Petruk: **242** (top). Stadtbibliothek im Bildungscampus Nürnberg, Amb. 317.2°, f. 34v: **67** (top). Stiftung der Werke von C.G. Jung (CC BY-SA 4.0): **121** (both). University of Glasgow Archives & Special Collections, Courtesy of; Ferguson Ah-y.9: **100**. University of Heidelberg; Photo: Elke Fuchs © Institute for Papyrology: **49**. University of Michigan Library: **57** (Digital Collections, PDM); **48, 50** (Papyrology Collection). University of Oslo Library; © Digital Corpus of Literary Papyri (CC BY 3.0): **47** (both). University of Toronto – Robarts Library: **97**. Wellcome Collection, London (PDM): **4, 8, 61, 88–89, 95, 102, 109, 120, 124** (both), **125, 127**. Wellcome Collection, London (CC BY 4.0): **130–131, 137**. "Wicked: The Grimmerie"; Designer: Headcase Design (Paul Kepple, Jude Buffum); Publisher: Hyperion, Melcher Media (packager); Author: David Cote: **234**. Württemberg State Museum/H. Zwietasch (CC-BY-SA 4.0): **6**. Zurich Central Library, Ms. F 23, p. 399: **106, 107, 108–109**.

Produced by
Quintessence Editions
1 Triptych Place, Second Floor
London SE1 9SH

Senior Editor: Emma Harverson
Design: Blok Graphic
Senior Art Editor: Rachel Cross
Picture Research: Sara Ayad
Associate Publisher: Eszter Karpati
Publisher: Lorraine Dickey

Published in association with
Yale University Press
302 Temple Street
P.O. Box 209040
New Haven, CT 06520-9040
yalebooks.com

Library of Congress Control Number: 2023935192
ISBN 978-0-300-27201-7

10 9 8 7 6 5 4 3 2 1

Printed in China

COVER: Title page from *De Alchimia: Opuscula complura veterum philosophorum* (*On Alchemy: Several Works of the Ancient Philosophers*), Frankfurt (1550). See page 120.